T0297545

EMBEDDED RTOS DESIGN

EMBEDDED RTOS DESIGN

Insights and Implementation

COLIN WALLS

Newnes is an imprint of Elsevier
The Boulevard, Langford Lane, Kidlington, Oxford OX5 1GB, United Kingdom
50 Hampshire Street, 5th Floor, Cambridge, MA 02139, United States

British Library Cataloguing-in-Publication Data
A catalogue record for this book is available from the British Library

Library of Congress Cataloging-in-Publication Data
A catalog record for this book is available from the Library of Congress

ISBN: 978-0-12-822851-7

For Information on all Newnes publications
visit our website at https://www.elsevier.com/books-and-journals

Publisher: Mara Conner
Acquisitions Editor: Tim Pitts
Editorial Project Manager: Mariana Henriques
Production Project Manager: Kamesh Ramajogi
Cover Designer: Mark Rogers

Typeset by MPS Limited, Chennai, India

Dedication

Dedicated to everybody who has taught me stuff: my parents, wives, children, and other family members; my teachers and those whom I have taught; colleagues, managers, and associates. I hope that I can give something back.

Contents

Foreword

Embedded systems have steadily become more complex with more devices than ever being connected to the Internet. The timing requirements for these systems have become more complicated as the system must meet all its scheduling deadlines for sampling sensors, managing Wi-Fi, Bluetooth, USB, interacting with the user through a GUI, securing communications through encryption, authenticating and verifying integrity through hashing and signatures along with many other common system activities. With so many activities competing for CPU time, developers must rely on a real-time operating system (RTOS) to help schedule all their tasks in order to ensure all their deadlines and timing requirements can be met.

Over the last decade, RTOS usage has increased dramatically to the point where an RTOS is used in nearly two-thirds of embedded systems projects! (Aspencore 2019 Embedded Market Survey) An RTOS has become a critical, foundational component that modern systems are built upon and the tools that an RTOS provides a developer are indispensable to ensuring that task deadlines are met and that system response times are reasonable.

Despite the wide use of RTOSes, many embedded developers do not have an in-depth understanding of what is happening under the hood with an RTOS. This book provides developers with the behind-the-scenes knowledge necessary to make critical decisions for their system architecture and implementation. Colin does a fantastic job not only explaining how various RTOS components are used but also explaining how they are built so that the reader can use this book to also build their own lightweight RTOS.

The book dives into the kernel and provides the reader with an in-depth understanding of how a commercial RTOS kernel works. This knowledge removes the veil and mystery that often surrounds using an RTOS in a product and provides the reader with an understanding that allows them to truly master RTOS design no matter what RTOS it is they are using. The book provides the reader with an example kernel that can be leveraged to then build their own RTOS features around. All the main elements of an RTOS have a chapter each to cover details of their operation and implementation.

One of the most important RTOS features in my opinion, and one that is the most neglected, can be found in Chapter 9, Event flag groups. Event flag groups are an incredibly efficient way to synchronize tasks within an RTOS-based application and as the reader might expect, they can be used to notify the system that specific events have occurred in the system such as button presses, over temperature alerts, and many other events. Developers often neglect event flag groups and choose to use semaphores that often use more code space, memory, and are less efficient from an execution standpoint. (More on semaphores can be found in Chapter 10: Semaphores.)

An RTOS is certainly not complete without additional synchronization and mutual exclusion facilities. Colin explains how to build and integrate RTOS components such as mailboxes (Chapter 11: Mailboxes), queues (Chapter 12: Queues), and pipes (Chapter 13: Pipes). These components help developers move data and messages around the application and can throw a kernel or application developer a context curveball if they are not careful with how they handle interrupts.

There are plenty of other topics in this book that the reader will find interesting. For example, Chapter 18, Diagnostics and error checking, goes into details on diagnostic and error checking facilities that are included in an RTOS. Diagnostic information can be absolutely critical to a developer successfully debugging their application and understanding how that application behaves. Diagnostic information can include as little as how many times a task executes in a given period all the way through a full-fledged event recorder that can be used to trace the RTOS execution and visualize how its application is performing.

Whether you are a developer looking to gain insights into how an RTOS works behind the scenes or are a kernel developer looking to jump start your own RTOS kernel, you are going to find that this book provides you everything you need to understand to develop with and build an RTOS.

Jacob Beningo
Embedded Software Consultant
www.beningo.com

Preface

Having worked with real-time operating systems (RTOSes) throughout my career—applying them, supporting users, and writing them myself—it was inevitable that I would write about them at some point. Indeed, over the years I have written countless articles, blog posts, and conference presentations on the topic. Even my first book, *Programming Dedicated Microprocessors*, which was published way back in 1986 and I often claim to be the first book ever on embedded software, had a chapter covering a simple task scheduler.

This book has quite a long history. Back in the 1990s, Accelerated Technology developed an RTOS product called Nucleus. This went on to be the most widely deployed operating system on the planet, with over 5 billion instances in operation. Accelerated Technology was acquired by Mentor Graphics, which, in turn, was acquired by Siemens, but Nucleus goes on.

Apart from being a very scalable, hard real-time operating system—which addressed the key technical requirements—Nucleus had an attractive business model: it was royalty-free and source code was supplied. This appealed to developers of high-volume products, as they had a very clear idea of costs while operating in markets where every cent would count. Developers felt comfortable that they were not using a "black box," as they could peruse the source code if need be; to my knowledge, very few customers ever do look at it, but are pleased to have it to hand "just in case."

We concluded that educating more engineers about using an RTOS (with the hope that they would eventually choose Nucleus of course) would be a good plan. We decided that someone (me) should write a new kernel in a "clean room" fashion—that is, without seeing the Nucleus product's source code—which was architecturally similar to Nucleus, but simplified. It should also be useable on smaller CPUs (16-bit or even 8-bit devices), but need not scale up for large applications. The code style should err on the side of simplicity and clarity. We dubbed the kernel "Nucleus SE" (where "SE" might mean "Starter Edition").

The plan was, rather than create a product to sell or give away, we (I) would write a book describing what an RTOS does, how it is used, and how it works, including the Nucleus SE code. This is that book.

Initially, after developing the kernel, I published a series of articles—*RTOS Revealed*—on embedded.com addressing this topic. I acknowledge the help of Steve Evanczuk with getting those articles out there. Those articles were the basis for this book.

I hope that this book provides useful guidance, as sharing knowledge and experience is, I believe, the professional way for any engineer to behave. Over the years, many colleagues—too many to mention—have guided me and shared their knowledge and I am grateful for that. I would like to reassure you that by buying this book you (or whoever did buy it) are not topping up my retirement fund or my employers' coffers, as they have generously permitted all royalties from sales of the book to be redirected to a charitable organization: LINC—www.lincfund.org—who provide invaluable support to sufferers of various forms of cancer and their families. My late wife contracted leukemia, which is how I encountered this organization and came to value their work.

Although the primary goal of the book is to give a clear idea of what an RTOS does and how to use it, the implementation details give some idea of how it does it. My intention is that this should not only consolidate understanding, but also provide the option of implementing RTOS functionality if needed. Through the book, all the code for Nucleus SE is included, with descriptions of its operation. There is no need to key in the code yourself, it is available for download. You can use the code in your own projects, if you wish—Nucleus SE is not a product, so there is no license, no support, and no warranty. You can, of course, contact me with any queries, suggestions, or feedback by e-mail (colin.walls@siemens.com or mail@colinwalls.com) or via social media.

Program structure and real time

This book is about embedded systems—specifically the software that runs in an embedded system. It is worth starting by making sure that we are all on the same page and have our terminology straight. So, what is an embedded system? When I first wrote a book on this topic—back in 1986—the word "embedded" did not occur in the title or anywhere in the text. This was simply because the term had not yet been coined. We were very much at a loss to give a handle to the systems we worked on, using terms like "dedicated systems" or "microsystems," none of which were very satisfactory. A few years later, the word embedded started to be used and was rapidly adopted by everyone in the field.

So, back to my question, what is an embedded system? Having had a lot of practice at explaining to friends and family about what I work on, I tend to say something like "any electronic device that contains a microprocessor (CPU) that would not normally be described as a computer."

An operating system (OS) is always used on a computer; the use of an OS of some kind on modern embedded systems is common. Although prevalent in high-end (32- and 64-bit mainly) systems, there may be benefits from using a kernel in lower power devices. OSs are very much the focus of this book, with lots of details in the chapters that follow—first about OSs in general, then taking a look at a specific implementation in depth.

Why use an operating system?

Having established that the key topic of this book is OSs in embedded applications, it is worth just checking why they are used. There are various explanations, some of which have got as much to do with human nature as they have technical requirements. I am reminded of a story. In one of our offices, that I used to visit, there was a kitchen where one could prepare coffee. On the door was a sign which said "Please do not close this door." Underneath, someone had written "Why not?" To

Embedded RTOS Design. DOI: https://doi.org/10.1016/B978-0-12-822851-7.00001-1

that, someone else had responded: "Cuz." This is, of course, an abbreviation for "Because," which, in turn, is short for "Because we are telling you to behave in this way." This is why an OS is used on some systems—just because that is what is done: cuz (Fig. 1.1).

Another explanation comes from looking at desktop applications. If you are going to write some software for a PC or a Mac, where do you start? Having got the computer, you switch it on and it starts up in Windows, Linux, or macOS and you start programming. The OS is a given, and provides useful services. It is very unlikely that you would consider starting from scratch, programming "bare" hardware. So, it is not surprising that an engineer, who has some software experience, but is new to embedded software, would expect the "safety blanket" of an OS on their embedded system.

It is worth noting that the key aspect of a desktop OS, that users are aware of, is the user interface (UI). Ask someone what Windows is all about and they will mention windows, menus, dialogs, and icons, but are less likely to talk about file systems and interprogram communication and interoperability. This is a fundamental difference between a desktop and an embedded system: an embedded device might not even have a UI—if it does, it may be quite simplistic. This is the first of

Figure 1.1 A common reason for doing anything.

many key differences that become apparent with a little thought:

- An embedded system normally runs a single software application from the moment it is switched on until it is shut down.
- Embedded systems have limited resources. Memory space may be large enough, but it is unlikely to be extensible; the CPU is probably just powerful enough, but with no extra capacity to spare.
- Many embedded applications are "real time." I will consider more carefully what this means later in this chapter.
- Embedded software development tools are quite specialized and run on a "host" computer (like a PC), not on the final ("target") system.
- The updating of embedded software in service is challenging. Although, with connected devices, possibilities are beginning to emerge, in-field updates are still not the norm (compared with the regular updates and patches applied to the desktop software).

As you will see, an embedded OS provides a useful programming model to the developer, which can make its use very attractive.

Software structure

Before looking at the ways that an embedded software application might be structured, it is worth thinking about the paradigms used on computers to facilitate the execution of programs under the control of an OS. I will do this by considering some of the approaches used in recent years.

First, there is "DOS style" program execution, where programs are executed in sequence (Fig. 1.2).

Here each program is started, utilized, and then terminated. We use Program 1, then Program 2, perhaps then take a break, return to use Program 3, and then come back to using Program 2 again. The second use of Program 2 is from scratch—it does not continue from where it left off earlier (unless this specific application provides that capability itself).

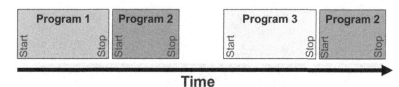

Figure 1.2 "DOS style" program execution.

After DOS, life became more complex in many ways, as Windows became the norm. With "Windows style" program execution, multiple programs may be running (apparently) concurrently (Fig. 1.3).

Here multiple programs appear to be running at the same time, with Windows managing that illusion. First, we start Program 1, then Program 2 is started as well, then Program 3. Program 2 finishes, leaving Programs 1 and 3 still running. Program 3 finishes leaving just Program 1. Later Program 2 is running again, and Program 1 terminates leaving just Program 2 running. This is quite a realistic scenario for the use of Windows by a typical user. The OS maintains the resources so that all the programs share the CPU in a reasonable way. It also facilitates easy communication between programs (the Clipboard is one example) and controls the UI.

There are some circumstances, notably with handheld devices, where more flexibility than DOS is required, but, with limited resources, lower overheads than Windows is necessary. This can result in "iOS style" program execution (Fig. 1.4).

In this situation, programs are run in sequence, but the state of each one is automatically normally saved so that it can continue from where it left off. In the diagram, we start off with Program 1, then this is paused to allow the use of Program 2, then perhaps the device is turned off for a while. It is restarted straight into Program 3 (Program 2's state had been automatically saved), and then the user returns to using Program 2 continuing their work

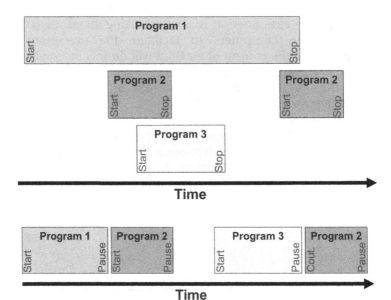

Figure 1.3 "Windows style" program execution.

Figure 1.4 "iOS style" program execution.

from earlier in the day. (I am aware that the iOS app execution model is rather more complex than this, but my description encapsulates the user's primary perspective.)

Most embedded applications do not really conform to any of these models, as typically the device starts running a program at power on and continues to run only that software indefinitely. The structure of such code needs to be considered carefully.

Embedded program models

Desktop systems are all much the same. From an application program's point of view, every Windows of PC is identical. Embedded systems are unique—everyone is different from every other one. These differences may simply be technical: the CPU choice, the amount of memory, and the range of peripheral devices may all vary. The priorities of the application program may differ: execution speed, power consumption, security, or reliability may be the critical factor(s). The differences may also be commercial, affecting the price: the production volume and the choice between custom or standard hardware are examples.

These differences have many implications for the embedded software developer. The choice of development tools (compilers, debuggers, and the like) is strongly affected by the choice of CPU, for example. Many factors affect the choice of an OS or even the decision to employ one. The software structure—the program model—needs to be carefully reviewed for every embedded software application.

Depending on the requirements of the application, embedded software may be structured in several ways, each of increasing complexity (Fig. 1.5).

At its simplest, an embedded application may be structured as just a loop, where it goes around repeatedly executing the same sequence of actions. If the application is simple enough to be implemented this way, it is an excellent choice—simple code is reliable and easy to understand. However, this structure is very sensitive to part of the code "hogging" the processor—that is, some instructions taking too long to run and delaying the execution of other parts of the application. This model is also not very scalable—enhancing the code may be challenging, as the additions may impact the performance of the older code.

If something a little more sophisticated is required, the next step is to consider placing all the nontime-critical code in the main loop and locate the time-sensitive parts in interrupt service routines (ISRs). The ISRs would typically be quite short,

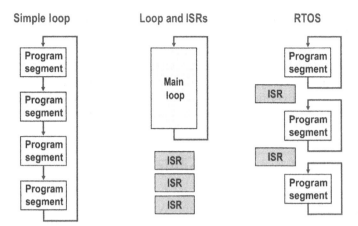

Figure 1.5 Embedded programming models.

performing only vital work and flagging part of the main loop to complete the job when it can. This is a good next step on the complexity ladder and again can be the right choice if the needs of the application are met. The challenge can come with the distribution of work between the main loop and the ISRs (and also between multiple developers).

For the ultimate in flexibility, a means is required to divide an application into a number of individuals, somewhat independent programs (which we will call tasks or threads), which are run in an apparently concurrent fashion. Small ISRs are also likely to be included in the system, but will mostly serve to notify tasks or trigger action. To achieve this, an OS—or at least a kernel—is required. The application of multithreading not only facilitates the flexible distribution of functionality within the software but also eases the apportioning of work across a development team.

What is real time?

Earlier I made the comment that many embedded applications are "real time." It is also common to talk in terms of a real-time OS (RTOS) in this context instead of simply an OS. So, now would be a good time to define our terms.

Here is one definition of a real-time system:

A real-time system is one in which the correctness of the computations not only depends on the logical correctness of the computation but also on the time at which the result is produced. If the timing constraints of the system are not met, system failure is said to have occurred.

The most important characteristic of such a system is predictability—or, as it is commonly termed, determinism.

A real-time system is not necessarily very fast—"real time" does not always mean "real fast." It means that any required action will be done in a timely manner. This means fast enough, but it can also mean not too fast (i.e., slow enough).

An RTOS can (if used the right way) provide very precise control over the allocation of CPU time to tasks and, hence, render an application entirely deterministic. But there is one factor that can get in the way of this ideal: interrupts. There are some RTOS products that control interrupts entirely. Their approach is to manage the servicing of the interrupt within the task scheduling scheme. Although this should lead to very predictable behavior, the mechanism tends to be rather complex and imposes a significant overhead. Most RTOSes simply allow ISRs to "steal" time from the task running at the moment that the interrupt occurs. This, in turn, places the onus on the programmer to keep the ISR code as short as possible. The result is a reasonable approximation of true real time. The only significant complication is catering for RTOS service calls within the ISR. Some calls are quite innocuous, while others will cause a context switch on return from interrupt; this situation needs to be specifically accommodated and this is done in various ways by different RTOSes.

2

Multitasking and scheduling

We have already considered the multitasking concept—multiple quasi-independent programs apparently running at the same time, under the control of an operating system. Before we look at tasks in more detail, we need to straighten out some more terminologies.

Tasks, threads, and processes

We use the word "task"—and I will continue to do so—but it does not have a very precise meaning. Two other terms—"thread" and "process"—are more specific and we should investigate what they mean and how they are differentiated.

Most often, a real time operating system (RTOS) used in an embedded application will employ a multithread model. Several threads may be running and they all share the same address space (Fig. 2.1).

This means that a context swap is primarily a change from one set of central processing unit (CPU) register values to another. This is quite simple and fast. A potential hazard is the ability of each thread to access memory belonging to the others or to the RTOS itself.

The alternative is the multiprocess model. If several processes are running, each one has its own address space and cannot access the memory associated with other processes or the RTOS (Fig. 2.2).

This makes the context swap more complex and time-consuming, as the OS needs to set up the memory management unit (MMU) appropriately. Of course, this architecture is only possible with a processor that supports an MMU. Processes are supported by "high end" RTOSes and most desktop operating systems. To further complicate matters, there may be support for multiple threads within each process. This latter capability is rarely exploited in conventional embedded applications.

A useful compromise may be reached, if an MMU is available (Fig. 2.3).

Embedded RTOS Design. DOI: https://doi.org/10.1016/B978-0-12-822851-7.00002-3

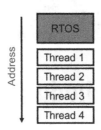

Figure 2.1 Thread model.

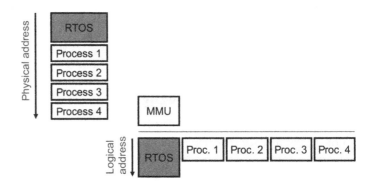

Figure 2.2 Process model.

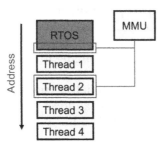

Figure 2.3 Thread protected mode.

Many thread-based RTOSes support the use of an MMU to simply protect memory from unauthorized access. So, while a task is in context, only its code/data and necessary parts of the RTOS are "visible"; all the other memory is disabled and an attempted access would cause an exception. This makes the context switch just a little more complex, but renders the application more secure. This may be called "Thread Protected Mode" or "Lightweight Process Model."

Schedulers

As we know, the illusion that all the tasks are running concurrently is achieved by allowing each to have a share of the processor time. This is the core functionality of a kernel. The way that time is allocated between tasks is termed "scheduling." The scheduler is the software that determines which task should be run next. The logic of the scheduler and the mechanism that determines when it should be run is the scheduling algorithm. We will look at several scheduling algorithms in this section. Task scheduling is actually a vast subject, with many whole books devoted to it. The intention here is to just give sufficient introduction that you can understand what a given RTOS has to offer in this respect.

Run to completion scheduler

Run to completion (RTC) scheduling is very simplistic and uses minimal resources. It is, therefore, an ideal choice, if the application's needs are fulfilled. Here is the timeline for a system using RTC scheduling (Fig. 2.4).

The scheduler simply calls the top-level function of each task in turn. That task has control of the CPU (interrupts aside) until the top-level function executes a return statement. If the RTOS supports task suspension, then any tasks that are currently suspended are not run. This is a topic discussed in the "*Task suspend*" section.

The big advantages of an RTC scheduler, aside from its simplicity, are the need for just a single stack and the portability of the code (as no assembly language is generally required). The downside is that a task can "hog" the CPU, so careful program design is required. Although each task is started "from the top" each time it is scheduled—unlike other kinds of schedulers which allow the code to continue from where it left off—greater flexibility may be programmed by use of static "state" variables, which determine the logic of each sequential call.

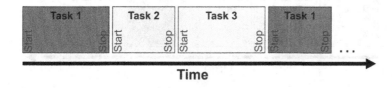

Figure 2.4 Run to completion scheduling.

Round robin scheduler

A round robin (RR) scheduler is similar to RTC, but more flexible and, hence, more complex. In the same way, each task is run in turn (allowing for task suspension) (Fig. 2.5).

However, with the RR scheduler, the task does not need to execute a return in the top-level function. It can relinquish the CPU at any time by making a call to the RTOS. This call results in the kernel saving the context (all the registers—including stack pointer and program counter) and loading the context of the next task to be run. With some RTOSes, the processor may be relinquished—and the task suspended—pending the availability of a kernel resource. This is more sophisticated, but the principle is the same.

The greater flexibility of the RR scheduler comes from the ability for the tasks to continue from where they left off without any accommodation in the application code. The price for this flexibility is more complex, less portable code, and the need for a separate stack for each task.

Time slice scheduler

A time slice (TS) scheduler is the next step in complexity from RR. Time is divided into "slots," with each task being allowed to execute for the duration of its slot (Fig. 2.6).

In addition to being able to relinquish the CPU voluntarily, a task is preempted by a scheduler call made from a clock tick

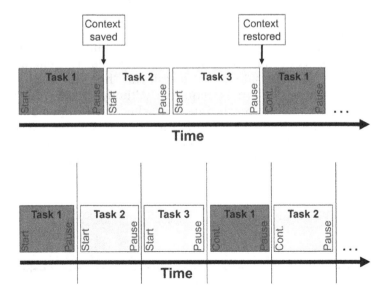

Figure 2.5 Round robin scheduling.

Figure 2.6 Time-sliced scheduling.

interrupt service routine (ISR). The idea of simply allocating each task a fixed time slice is very appealing—for applications where it fits the requirements—as it is easy to understand and very predictable.

The only downside of simple TS scheduling is the proportion of CPU time allocated to each task varies, depending on whether other tasks are suspended or relinquish part of their slots (Fig. 2.7).

In this example, Task 2 suspends itself, thus Task 3, then Task 1 are scheduled early. Because Task 2 is suspended, the next scheduling of Task 3 is very early.

A more predictable TS scheduler can be constructed if the concept of a "background" task is introduced. The idea, shown here, is for the background task to be run instead of any suspended tasks and to be allocated the remaining slot time when a task relinquishes (or suspends itself) (Fig. 2.8).

Obviously, the background task should not do any time-critical work, as the amount of CPU time, it is allocated as totally unpredictable—it may never be scheduled at all.

This design means that each task can predict when it will be scheduled again. For example, if you have 10 ms slots and 10 tasks, a task knows that, if it relinquishes, it will continue executing after 100 ms. This can lead to elegant timing loops in the application tasks.

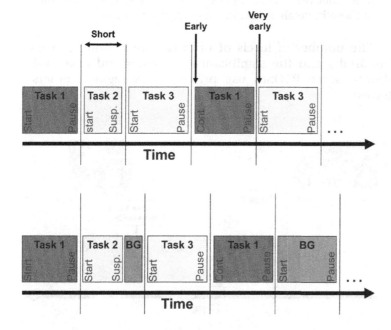

Figure 2.7 A problem with time-sliced scheduling.

Figure 2.8 Time-sliced scheduling with a background task.

An RTOS may offer the possibility for different time slots for each task. This offers greater flexibility but is just as predictable as with fixed slot size. Another possibility is to allocate more than one slot to the same task, if you want to increase its proportion of the allocated processor time.

Priority scheduler

Most RTOSes support priority scheduling. The idea is simple: each task is allocated a priority and, at any particular time, whichever task has the highest priority and is "ready" allocated to the CPU (Fig. 2.9).

The scheduler is run when any "event" occurs (e.g., interrupt or certain kernel service calls) that may cause a higher priority task being made "ready." There are broadly three circumstances that might result in the scheduler being run:

• The task suspends itself; clearly the scheduler is required to determine which task to run next.
• The task readies another task (by means of an application programming interface [API] call) of higher priority.
• An ISR readies another task of higher priority. This could be an input/output device ISR or it may be the result of the expiration of a timer (which are supported by many RTOSes—we will look at them in detail in Chapter 15—*Application timers*).

The number of levels of priority varies (from 8 to many hundreds) and the significance of higher and lower values differs; some RTOSes use priority 0 as highest, others as lowest.

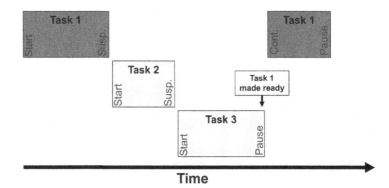

Figure 2.9 Priority scheduling.

Some RTOSes only allow a single task at each priority level; others permit multiple tasks at each level, which complicates the associated data structures considerably. Many OSes allow task priorities to be changed at runtime, which adds further complexity.

Composite scheduler

We have looked at RTC, RR, TS, and priority schedulers, but many commercial RTOS products offer more sophisticated schedulers, which have characteristics of more than one of these algorithms. For example, an RTOS may support multiple tasks at each priority level and then use time slicing to divide time between multiple ready tasks at the highest level.

Task states

At any one moment in time, just one task is actually running. Aside from CPU time spent running ISRs (more on that in *Interrupts* later in this chapter) or the scheduler, the "current" task is the one whose code is currently being executed and whose data are characterized by the current register values. There may be other tasks that are *Ready* (to run) and these will be considered when the scheduler is executed. In a simple RTOS, using an RTC, RR, or TS scheduler, this may be the whole story. But, more commonly, and always with a priority scheduler, tasks may also be in a *Suspended* state, which means that they are not considered by the scheduler until they are resumed and made *Ready*.

Task suspend

Task suspension may be quite simple—a task suspends itself (by making an API call) or another task suspends it. Another API call needs to be made by another task or ISR to resume the suspended task. This is an "unconditional" or "pure" suspend. Some OSes refer to a task as being "asleep."

An RTOS may offer the facility for a task to suspend itself (go to sleep) for a specific period of time, at the end of which it is resumed (by the system clock ISR). This may be termed "sleep suspend."

Another more complex suspend may be offered, if an RTOS supports "blocking" API calls. Such a call permits the task to request a service or resource, which it will receive immediately if it is available, otherwise it is suspended until it is available.

There may also be a timeout option whereby a task is resumed if the resource is not available in a specific timeframe.

Other task states

Many RTOSes support other task states, but the definition of these and the terminology used varies. Possibilities include a *Finished* state, which simply means that the task's outermost function has exited (either by executing a return or by just ending the outer function block). For a *Finished* task to run again, it would probably need to be reset in some way.

Another possibility is a *Terminated* state. This is like a pure suspend, except that the task must be reset to its initial state in order to run again.

If an RTOS supports dynamic creation and deletion of tasks (see later in this chapter), this implies another possible task state: "deleted."

Task identifiers

Clearly, it is necessary to be able to identify and specify each task in a system. This is, of course, a requirement for other kernel objects (discussed in future chapters) as well, but tasks have some subtle nuances that make a discussion of the broader topic appropriate here.

The designers of different RTOSes have taken different approaches to task identifiers, but four broad strategies can be identified:

1. A task is identified by means of a pointer to its "control block." This approach seems sensible; such a pointer would always be unique to a specific task and is useful, as access to the control block is required by many API calls. The downside is that this implies that all the data about a task are held in contiguous memory (RAM), which may be inefficient. A pointer is also likely to require 32 bits of storage.
2. A task may be identified by an arbitrary "index number." This value may be useful by giving access to entries in specific tables. Such an identifier may only require eight bits or less of storage, depending on any limitations on the number of tasks that are supported by the RTOS.
3. Some RTOSes only allow one task per priority level and, hence, use the priority to uniquely identify the task. This

implies that a task's priority may not be changed. This approach is really a variation of (2).

4. Tasks might have names which are character strings. This may be useful for debugging, but is unlikely to be an efficient means of uniquely identifying a task otherwise. RTOSes that support task naming generally have an additional identifier (like a pointer), which is used by API calls, etc. For most embedded systems, textual names are an unnecessary overhead on the target; a good debugger would allow naming locally on the host.

The context switch

When the scheduler changes which task is currently running, this is termed a "context switch." This is a matter that is worth closer study, as the way that a context switch works is fundamental to an RTOS design.

What is a task?

We know that a task is a quasi-independent program, which shares CPU time with several other tasks under the control of an RTOS. But we need to think about what really characterizes a task.

Register set

A task is ultimately a unique set of processor register values. These are either loaded into the CPU registers (i.e., the task is current) or stored somewhere ready to be loaded when the task is scheduled. In an ideal world, a CPU would have multiple sets of registers, such that one may be allocated to each task. This has been realized in some special cases. Many years ago, the Texas TI9900 series had multiple, per-task register sets, but these were implemented in main memory, which yielded limited performance. The SPARC architecture—well known historically for its use in Unix desktop systems—has a facility for maintaining multiple register sets in a ring structure, but the total number available is rather limited.

Local data

A task will probably have its own stack, the size of which may be set on a task-by-task basis or may be a global setting for all tasks in the system. This, along with the registers, provides task-specific data storage. There may be other data storage memory areas that are dedicated to a specific task.

Shared resources

Just about everything else may be shared between tasks.

Code may commonly be shared—either certain functions or the complete code for a task. Care is needed to ensure that the code is reentrant—primarily no static variables (either explicitly declared static or simply declared outside of functions) should be used. Beware of the standard library modules that are not designed for embedded use; there are commonly many non-reentrant functions.

Sharing data are also possible, but care is needed to ensure that access to the data is tightly controlled. Ideally just one task is "owner" of the data at any one time.

Saving context

When a task is descheduled (i.e., ceases to be current), its set of registers needs to be saved somewhere. There are at least two possibilities:

- The registers may be stored in a task-specific table. This may be part of a task control block (TCB). The size is predictable and constant (for a specific CPU architecture).
- The registers may be pushed onto the task's stack. This necessitates the allocation of sufficient extra stack space and provision of storage for the stack pointer (in the TCB, possibly).

Which mechanism is chosen depends on the design of a particular RTOS and on the target processor. Some (typically 32-bit) devices can address stack very efficiently; others (8-bit for example) may be better with tables.

Dynamic task creation

A fundamental aspect of the architecture of an RTOS is whether it is "static" or "dynamic."

With a static RTOS, everything is defined at application build time—notably the number of tasks in the system. This seems reasonably logical for embedded applications, which typically have a specific, bounded functionality.

A dynamic RTOS, on the other hand, starts up with a single task (which may be a special "master" task or suchlike) and creates and deletes other tasks as and when required. This enables the system to be adaptable to changing requirements and is a much closer analog to a desktop system, which behaves in exactly this fashion.

The static/dynamic characterization also applies to other kernel objects, which will be discussed in later chapters.

Requirements for dynamic task creation

This capability is offered by most commercial RTOS products. However, only a small proportion of applications really have a need for the dynamic behavior. It is very common for a system to start, create all the necessary tasks (and other objects) then never create or destroy any more during the running of the application code. The dynamic task creation functionality is so ubiquitous simply because it became a "checkbox item." One RTOS vendor introduced it, and all the others followed suit.

Notably, the OSEK/VDX standard demands a static architecture—even though this may be applied to quite complex applications. A result of this requirement is the inability to implement OSEK/VDX by means of an intermediate layer—a "wrapper"—on top of a conventional (dynamic) RTOS product.

Implications of dynamic task creation

There are several issues associated with dynamic behavior, which may be of concern.

First, there is the introduction of complexity, which means that the data structures that describe tasks (TCBs) need additional information; generally, they are implemented as doubly linked lists, which adds a two-pointer overhead to the memory footprint.

All the data that describe a task must be held in RAM. This is inefficient, as much of it may just be constant data items copied from ROM. Also, on low-end processors (microcontrollers) RAM may be in short supply.

Probably the biggest worry is the potential for unpredictable resource shortages, leading to the failure to create new objects. Since the crux of a real-time system is its predictability, this sounds like an unacceptable situation. So, care is needed in the use of dynamic task (and other objects) creation.

Interrupts

Although it is conceivable that a real-time embedded system might be implemented without the use of interrupts, it would be most unusual.

Interrupts and the kernel

When an RTOS is in use, an ISR is normally implemented to be as lightweight as possible—to "steal" the minimum amount of CPU time from the scheduled tasks. Often, a device may simply be serviced and any required work be queued up ready for processing by a task. Beyond this, it is hard to generalize about interrupts and their relationship to kernels simply because there is a lot of variabilities. At one extreme, the RTOS designer may conclude that interrupts are not the business of the kernel at all and it is left to the programmer to be careful not to compromise task scheduling by using too much CPU time in ISRs. At the other end of the scale, an RTOS may take complete control of the entire interrupt subsystem. Neither approach is right or wrong; they are just different.

Context save

An ISR always needs to save the "context" so that the interrupted code is unaffected by the computations of the ISR. In a system implemented without an RTOS, this is simply a matter of preserving any registers used by the ISR—generally on the stack—and restoring them before return. Some processors have provision for a dedicated ISR stack—others simply use the same one as the application code.

When an RTOS is in use, the approach may be exactly the same. In the same fashion, the stack used by the ISR may be "borrowed" from the currently running task or it may be another stack dedicated to interrupts. Some kernels implement this capability even if the CPU itself does not facilitate an interrupt stack. The situation becomes more complex if the ISR makes an API call which affects the scheduler. This may result in the interrupt returning to a different task from the one that was running when the interrupt occurred.

Interrupts and the scheduler

There are several circumstances when the ISR code operation may result in a return being made to a different task:

- With a priority scheduler, the ISR may have made ready a task with a higher priority than the current one.
- The ISR might suspend the current task.
- With a TS scheduler, the clock ISR will manage the TSs and may call the scheduler when required. More on this is explained below.

Tick clock

It is very common for an embedded system to have a periodic "tick clock." Indeed, some RTOS implementations mandate such a facility being available. Commonly though, the availability of a tick clock is optional and its absence simply precludes the availability of certain facilities. The tick clock ISR may typically provide four types of functionality:

- If a TS (or composite) scheduler is in use, the clock ISR will manage the tick count and schedule a new task each time it expires.
- Commonly "system time" is maintained. This is generally just a 32-bit variable that is incremented by the clock ISR and may be set up or queried by tasks.
- If the RTOS supports application timers, they will be maintained by the clock ISR, which would deal with any expiration or rescheduling requirements.
- If an RTOS supports timeouts on blocking API calls or tasks can be "asleep," these timings will be managed by the clock ISR.

3

RTOS services and facilities

So far, I have talked about a real-time operating system (RTOS) providing support for the multitasking model. This is the most fundamental capability of any kind of operating system (OS), but there is typically much more functionality provided in the form of services. These are normally accessed by means of an application program interface (API): a library of function calls that may be used by application code. The actual range of options offered by a different RTOSes may vary quite widely—as will some of the terminology—so the best I can do in this chapter is to review the commonly available facilities. Most of the rest of the book is dedicated to further details of RTOS services.

Intertask communication services

I have seen a task described as a quasi-independent program. Although tasks in an embedded application have a degree of independence, it does not mean that they have no "awareness" of one another. It is possible that some tasks will be truly isolated from others, but the requirement for communication and synchronization between tasks is very common and most RTOSes provide a variety of services that provide communication between, and synchronization of, tasks.

There are three broad paradigms for intertask communications and synchronization:

Task-owned facilities—attributes that an RTOS imparts to tasks that provide communication (input) facilities. The example that I will look at some more is signals.

Kernel objects—facilities provided by the RTOS which represents stand-alone communication or synchronization facilities. Examples include: event flags, mailboxes, queues/pipes, semaphores, and mutexes.

Message passing—a rationalized scheme where an RTOS allows the creation of message objects, which may be sent from one to task to another or to several others. This is fundamental to the kernel design and leads to the description of such a product as being a "message passing RTOS."

Embedded RTOS Design. DOI: https://doi.org/10.1016/B978-0-12-822851-7.00003-5

The facilities that are ideal for each application will vary. There is also some overlap in their capabilities and some thought about scalability is worthwhile. For example, if an application needs several queues, but just a single mailbox, it may be more efficient to realize the mailbox with a single-entry queue. This object will be slightly nonoptimal, but all the mailbox handling code will not be included in the application and, hence, scalability will reduce the RTOS memory footprint.

Shared variables or memory areas

A simplistic approach to intertask communication is to just have variables or memory areas which are accessible to all the tasks concerned. Whilst it is very primitive, this approach may be applicable to some applications. There is a need to control access. If the variable is simply a byte, then a write or a read to it will probably be an "atomic" (i.e., uninterruptible) operation, but care is needed if the processor allows other operations on bytes of memory, as they may be interruptible and a timing problem could result. One way to effect a lock/unlock is simply to disable interrupts for a short time.

If you are using a memory area, of course you still need locking. Using the first byte as a locking flag is a possibility, assuming that the memory architecture facilitates atomic access to this byte. One task loads data into the memory area, sets the flag, and then waits for it to clear. The other task waits for the flag to be set, reads the data, and clears the flag. Using interrupt disable as a lock is less wise, as moving the whole buffer of data may take time.

This type of shared memory usage is similar in style to the way many interprocessor communication facilities are implemented in multicore systems. In some cases, a hardware lock and/or an interrupt are incorporated into the interprocessor shared memory interface.

Signals

Signals are probably the simplest intertask communication facility offered in conventional RTOSes. They consist of a set of bit flags—there may be 8, 16, or 32, depending on the specific implementation—which is associated with a specific task.

A signal flag (or several flags) may be set by any task using an OR type of operation. Only the task that owns the signals can read them. The reading process is generally destructive—that is, the flags are also cleared.

In some systems, signals are implemented in a more sophisticated way such that a special function—nominated by the signal owning task—is automatically executed when any signal flags are set. This removes the necessity for the task to monitor the flags itself. This is somewhat analogous to an interrupt service routine.

There is more information about signals in Chapter 8, *Signals*, which describes their implementation in Nucleus SE.

Event flag groups

Event flag groups are like signals in that they are a bit-oriented intertask communication facility. They may similarly be implemented in groups of 8, 16, or 32 bits. They differ from signals in being independent kernel objects; they do not "belong" to any specific task.

Any task may set and clear event flags using OR and AND operations. Likewise, any task may interrogate event flags using the same kind of operation. In many RTOSes, it is possible to make a blocking API call on an event flag combination; this means that a task may be suspended until a specific combination of event flags has been set. There may also be a "consume" option available, when interrogating event flags, such that all read flags are cleared.

There is more information about event flag groups in Chapter 9, *Event Flag Groups*, which describes their implementation in Nucleus SE.

Semaphores

Semaphores are independent kernel objects, which provide a flagging mechanism that is very commonly used to control access to a resource. There are broadly two types: binary semaphores (that just have two states) and counting semaphores (that have an arbitrary number of states). Some processors support (atomic) instructions that facilitate the easy implementation of binary semaphores. Binary semaphores may also be viewed as counting semaphores with a count limit of 1.

Any task may attempt to obtain a semaphore in order to gain access to a resource. If the current semaphore value is greater than 0, the obtain will be successful, which decrements the semaphore value. In many OSes, it is possible to make a blocking call to obtain a semaphore; this means that a task may be suspended until the semaphore is released by another task. Any task may release a semaphore, which increments its value.

There is more information about semaphores in Chapter 10, *Semaphores*, which describes their implementation in Nucleus SE.

Mailboxes

Mailboxes are independent kernel objects, which provide a means for tasks to transfer messages. The message size depends on the implementation, but will normally be fixed. One to four pointer-sized items are typical message sizes. Commonly, a pointer to some more complex data is sent via a mailbox. Some kernels implement mailboxes so that the data is just stored in a regular variable and the kernel manages access to it. Mailboxes may also be called "exchanges," though this name is now uncommon.

Any task may send to a mailbox, which is then full. If a task then tries to send a full mailbox, it will receive an error response. In many RTOSes, it is possible to make a blocking call to send to a mailbox; this means that a task may be suspended, if the mailbox is full, until it is read by another task. Any task may read from a mailbox, which renders it empty again. If a task tries to read from an empty mailbox, it will receive an error response. In many RTOSes, it is possible to make a blocking call to read from a mailbox; this means that a task may be suspended, if the mailbox is empty, until it is filled by another task.

Some RTOSes support a "broadcast" feature. This enables a message to be sent to all the tasks that are currently suspended on reading a specific mailbox.

Certain RTOSes do not support mailboxes at all. The recommendation is to use a single-entry queue (see below) instead. This is functionally equivalent, but carries additional memory and runtime overhead.

There is more information about mailboxes in Chapter 11, *Mailboxes*, which describes their implementation in Nucleus SE.

Queues

Queues are independent kernel objects, which provide a means for tasks to transfer messages. They are a little more flexible and complex than mailboxes. The message size depends on the implementation, but will normally be a fixed size and word/ pointer oriented.

Any task may send to a queue and this may occur repeatedly until the queue is full, after which time any attempts to send will result in an error. The depth of the queue is generally user specified when it is created or the system is configured. In many RTOSes, it is possible to make a blocking call to send to a queue; this means that, if the queue is full, a task may be suspended until the queue is read by another task. Any task may read from a queue. Messages are read in the same order as they were sent—first in, first out (FIFO). If a task tries to read from an empty queue, it will receive an error response. In many RTOSes, it is possible to make a blocking call to read from a queue; this means that, if the queue is empty, a task may be suspended until a message is sent to the queue by another task.

An RTOS will probably support the facility to send a message to the front of the queue—this is also termed "jamming." Some RTOSes also support a "broadcast" feature. This enables a message to be sent to all the tasks that are suspended on reading a queue. Additionally, an RTOS may support the sending and reading of messages of variable length; this gives greater flexibility, but carries some extra overhead.

Many RTOSes support another kernel object type called "pipes." A pipe is essentially identical to a queue, but processes byte-oriented data.

The internal operation of queues is not of interest here, but it should be understood that they have more overheads in memory and runtime than mailboxes. This is primarily because two pointers—to the head and tail of the queue—need to be maintained.

There is more information about queues and pipes in Chapter 12, *Queues*, and Chapter 13, *Pipes*, which describe their implementation in Nucleus SE.

Mutexes

Mutual exclusion semaphores—mutexes—are independent kernel objects, which behave in a very similar way to normal binary semaphores. They are slightly more complex and incorporate the concept of temporary ownership (of the resource, access to which is being controlled). If a task obtains a mutex, only that same task can release it again—the mutex (and, hence, the resource) is temporarily owned by the task.

Mutexes are not provided by all RTOSes, but it is quite straightforward to adapt a regular binary semaphore. It would be necessary to write a "mutex obtain" function, which obtains the semaphore and notes the task identifier. Then a complementary "mutex release" function would check the calling task's identifier and release the semaphore only if it matches the stored value, otherwise it would return an error.

Other RTOS services

Although intertask communication and synchronization are important, an RTOS is likely to offer a great many other services of use to application code.

Task control

Beyond task scheduling and communication, an RTOS will undoubtedly include functionality—API calls—to control tasks in a selection of ways. We can look at some of the possibilities:

Task creation and deletion—In a "dynamic" RTOS, there will be calls to enable the creation of tasks (and other RTOS objects), as and when they are required. Such calls include a wide range of parameters which define the task; for example: code entry point, stack size, and priority. A corresponding task deletion API call enables resources to be released after the task is no longer required.

In a "static" RTOS, the defining parameters of a task are set up in a configuration file at build time.

Task suspend and resume—As we have seen, most RTOSes have the concept of a *Suspended* state for tasks. This may be entered in a variety of ways. One is for an explicit call to a Task Suspend API function. This may be called by the task itself or by another task. A corresponding Resume Task call enables the task to be a candidate for scheduling again.

Task sleep—In a real-time system, the control of time is obviously a requirement and it can take many forms. A simple one is the ability for a task to "sleep"—that is, the task is suspended for a specific period of time. When that time has elapsed, the task is "woken" and is a candidate for scheduling again. An API call will generally be available for this purpose. Of course, this functionality is dependent on the availability of a real-time clock.

Relinquish—When using a round-robin scheduler, a task may wish to give up control of the processor to the next task in the chain. A Task Relinquish API will be available for this purpose. The task is not suspended—it is available for scheduling when its turn comes around again. With a time-sliced scheduler, it may equally be desirable for a task to relinquish the remainder of its time slot, if it has no further useful work to do immediately. A task relinquish has no logical meaning with a Run to Completion or Priority scheduler.

Task termination—As we saw in Chapter 2, *Multitasking and Scheduling*, in addition to being *Ready* or *Suspended*, an RTOS may support other task states. A task may be *Finished*, which means its main function has simply exited—no special API call is required. A task may be *Terminated*, which means that it is not available for scheduling and must be reset in order to be available to run again—see "*Task reset*." This requires a specific API call. The availability of these additional task states, the terminology used, and their precise definitions will vary widely from one RTOS to another.

Task reset—Many RTOSes offer a Task Reset API call, which enables a task to be returned to its start-up condition. It may be in a *Suspended* state and require a Task Resume in order to become a scheduling candidate.

Task priority, etc.—In a dynamic RTOS, API calls are likely to be available to adjust several task characteristics at run-time. Examples include priority and time-slice duration.

There will be more information about task control in Chapter 6, *Tasks*, when we will look at its implementation.

System information

An RTOS will provide a range of API calls to provide tasks with information about the system, including:

Task information—how many tasks are in the system and details of their configuration and current status.

Information about other kernel objects—how many of each type of object is in the system and object-specific configuration and status information. For example:

- What is the current capacity of a queue to accept more messages?
- How many tasks are suspended on a specific mailbox?

RTOS version information. It is common for an API call to be available that returns such data.

Memory allocation

In many applications, it is useful for a program to be able to dynamically grab some memory when it is required, and release it again when it is no longer needed. Embedded software is no exception. However, the conventional approaches are prone to problems, which in desktop applications are unlikely or just inconvenient, but may be disastrous in an embedded system. There are ways to provide this kind of facility, even in a static RTOS.

Problems with malloc() and free()

In a desktop C program, a function can call `malloc()`, specifying how much memory is required and receive back a pointer to the storage area. Having made use of the memory, it may be relinquished with a call to `free()`. The memory is allocated from an area called the "heap." The problem with this approach is that, with an uncoordinated sequence of calls to these functions, the heap may easily become fragmented and memory allocations will then fail, even if enough memory is available, because contiguous areas are not large enough. In some systems (like Java and Visual Basic), elaborate "garbage collection" schemes are employed to effect defragmentation. The problems with these schemes are that it can result in very significant, unpredictable runtime delays and the need to use indirect pointers (which is not the way that C works). If `malloc()` and `free()` were implemented in a reentrant way (they usually are not) and used by RTOS tasks, the fragmentation would occur very rapidly and system failure would be almost inevitable.

In C++, there are operators, `new` and `delete`, which perform broadly the same functions as `malloc()` and `free()`. They also suffer from the same limitations and problems.

Memory partitions

A solution to providing dynamically available memory for a real-time system, in a reliable and predictable fashion, is to use a memory block-based approach. Such blocks are commonly called "partitions"; partitions may be allocated from a "partition pool."

A partition pool contains a certain number of blocks, each of the same size. The number and size of the blocks in a partition is determined when the partition pool is created. This may be dynamically, if the RTOS permits, or statically at build time. Typically, an application may include several partition pools offering blocks of various sizes.

If a task requires some memory, it makes an API call requesting a block from a specific pool. If that call is successful, the task will receive a pointer to the allocated block. If the call fails, because no partitions are currently available in the specified pool, the task may receive an error response. Alternatively, the task may be blocked (suspended) until another task frees a block in the partition.

It is usual for a task to simply pass on the pointer to the memory block to any code that needs to utilize it. This results in a problem when the block is no longer needed. If the code only has a pointer to the block, how can it tell the RTOS, in an API call, from which partition pool it wishes to free memory? The answer is that most RTOSes maintain extra data within the allocated block (typically a negative offset from the pointer) that provides the required information. Thus, the API call to free a block only requires its address.

There will be more information about memory partitions in Chapter 7, *Partition Memory*, which describes their implementation in Nucleus SE.

Time

It is self-evident that functionality concerned with the use and control of time are likely to be available in an RTOS. Exactly what is on offer will vary from one RTOS to another, but we can look at some common facilities. In all cases, a real-time clock is a given for any of these services to function.

System time—A simple system time, or "tick clock," is nearly always available. This is simply a counter (generally 32 bits), which is incremented by the real-time clock interrupt service routine, and may be set and read via API calls.

Service call timeouts—It is common for an RTOS to allow blocking API calls—that is, the calling task is suspended (blocked) until the requested service can be provided. Normally that blocking is indefinite, but some RTOSes offer a timeout, whereby the call returns when the timeout expires, if the service continues to be unavailable. API call timeouts are not supported in all RTOSes.

Task sleep—Tasks normally have an option to suspend themselves for a fixed time period. This was discussed earlier under *Task Control*.

Application timers—To enable application tasks to perform timing functions, most RTOSes offer timer objects. These are independent timers, updated by the real-time clock interrupt service routine (ISR), which may be controlled by API calls. Such calls configure, control, and monitor the timer operation. Typically, they may be set for one-shot or automatic restart operation. It is also common for an expiration routine to be supported—a function which is executed each time the timer completes a cycle.

There will be more information about system time and application timers in Chapter 14, *System Time*, and Chapter 15, *Application Timers*, which describe their implementation.

Interrupts, drivers, and input/output

The extent to which various RTOSes are concerned with interrupts and input/output is very variable. Likewise, some RTOSes have a very clearly defined structure for device drivers, which may be quite complex and be an issue when selecting a specific product.

Interrupts

Interrupts are an issue for an RTOS for two reasons:

- Without some safeguards, an ISR will "steal" CPU time, thus compromising the real-time behavior of the RTOS.
- If an ISR makes API calls which affect the task scheduling situation, this must be controlled and the RTOS must have the opportunity to run its scheduling algorithm. An example of such an API call would be one that wakes up a task of higher priority than the one that was running when the interrupt occurred.

Some RTOSes take complete control of all interrupts. A series of API calls are available to allow application ISRs to be "registered." This approach enables the scheduler to determine exactly when interrupts are allowed and facilitates the use of the majority of API calls from within an ISR.

Nucleus RTOS, for example, has a "low level" and "high level" ISR concept, which enables reliable interrupt control without introducing unnecessary overheads (i.e., increased interrupt latency).

Other RTOSes may take a somewhat "hands-off" approach to interrupts, leaving it to the applications programmer to ensure that ISRs are well-behaved. Typically, an additional ISR prefix and suffix are provided to secure API calls made therein.

Nucleus SE takes quite a light-weight approach to interrupt handling, which will be described in Chapter 16, *Interrupts in Nucleus SE.*

Drivers

Most RTOSes define some kind of device driver structure. The details vary from one RTOS to another, but a driver generally consists of two interacting components: some inline code (API calls) and an ISR. Typically other API calls will be available to control and register drivers.

Input/output

The majority of RTOSes currently on the market are not concerned with higher-level input/output, but some do define stream I/O, which is basically just establishing a connection between appropriate device drivers and standard C functions like `printf()`.

Historically, RTOSes would often support a "console"—a user interface to the RTOS via a serial line. This was largely for diagnostic and debug purposes. The use of modern, RTOS-aware debuggers has mostly removed the need for such facilities.

Diagnostics

RTOSes are usually required to minimize their memory footprint and offer greatest performance. Thus, integrity checking is not a high priority. Using modern, RTOS-aware debug technology, much checking can be done outside of the RTOS itself, but some facilities are also commonly implemented.

API call parameter checks

The parameters to RTOS API calls may be numerous and complex. Hence, errors are always a possibility. Many RTOSes offer runtime parameter checking, returning an error code if a parameter is invalid. Since this involves extra code and runtime performance is adversely affected by such checks, parameter checking is generally a build or configuration option.

Stack checking

For most types of scheduler (other than run to completion), each task has its own stack, the size of which is most commonly set on an individual basis. In some RTOSes, there is a separate stack for the kernel; in others, the task stack is "borrowed" during an API call. The integrity of stacks is clearly important to the overall reliability of the system. So, it is common for RTOSes to provide some provision for runtime integrity checking. There are a couple of possibilities:

- There may be an API call that returns the amount of stack space remaining for the current or specified task.
- Stacks may be configured with "guard words" at their base. These are set to a unique (normally odd, nonzero) value and checked for overwriting from time to time.

Application diagnostics

Although not generally supported directly by an RTOS, an application task may be dedicated to verifying the integrity of the whole system. Such a task may be responsible for the servicing of a hardware watchdog. The task may receive periodic input (e.g., a signal setting) from each of the critical tasks. Only when it has confirmed that all tasks have "checked in" does it service the watchdog and prevent a system reset.

Beyond the kernel

An RTOS is rather more than just a kernel, which is where I have focused attention thus far. This is definitely an area where a desktop OS differs widely from an embedded RTOS. On the desktop, all the additional components will normally be present and installed—all desktop machines have a graphical user interface and very few are not networked in some way, for example. A desktop machine has no real resource limitations; memory, hard drive space, and CPU power are available in abundance. In the resource-limited world of embedded systems, such additional components as graphics, networking, and a file system may be required, but they must be options and each should be scalable to minimize their memory footprint.

Networking and connectivity

Most embedded systems are "connected" in some way. It is, therefore, unsurprising that there is a lot of interest in embedded networking solutions and many products on the market.

TCP/IP—The standard protocol for the Internet is TCP/IP, which results in its very wide use, and makes it the obvious choice for many other applications. Typically, TCP/IP is used over Ethernet (IEEE802.3), which traditionally performs at 10 Mbps; nowadays 100 Mbps is quite common and 1 Gbps is coming along. It is also possible to use TCP/IP over other media. For example, Point-to-Point Protocol (PPP) is an implementation of TCP/IP over a serial line, which has been adapted in recent years for broadband internet connections.

The version of the fundamental IP protocol used until recently has been v4 (IPv4). However, this version is becoming outdated, as it is running out of address capacity. The solution is IPv6, which increases address capacity drastically and includes improved security and maintenance facilities. IPv6 is widely available and is demanded for equipment in most countries and for military applications everywhere.

An alternative, related protocol, User Datagram Protocol (UDP), is useful when maximum performance is required. UDP does not offer the reliability and ordering of TCP, but is light-weight and efficient.

USB—Universal Serial Bus is very widely used for devices that require connection to desktop computers. It offers a very easy to use, "plug & play" interface, which is the visible face of some quite complex software technology inside. An embedded device, that is to be connected to a PC, needs to be implemented as a USB "function," which requires a specific set of software components. If a device is intended to control other USB interfaced equipment (like a PC normally does), it will require a USB "host" software stack. A variant of USB—USB On-The-Go—enables a device to swap between being a host and a function dynamically.

IEEE1394—Another serial interface standard, which is used particular when large volumes of data need to be transferred between devices quickly (e.g., video data) is IEEE1394—also known as FireWire and i.Link.

Wireless—The convenience and popularity of various wireless technologies among consumers has led to the demand for such capabilities in embedded devices. Wi-Fi (the IEEE802.11 series) is true networking, allowing both peer to peer and infrastructure topologies over a reasonable range. There is increasing interest in the security of data in such networks, which must be addressed in the software. Other radio technologies—notably Bluetooth and ZigBee—provide short-range, point-to-point wireless connectivity.

Protocol validation

Since networking is very widely demanded, there are many vendors offering solutions. A problem for purchasers is verifying the quality of available products. Unlike an RTOS kernel, fully testing the functionality and performance of a protocol stack is a nontrivial process. Fortunately, validation suites are available—albeit at considerable cost—and the potential buyer can easily ask a vendor which suite(s) they are utilizing.

Graphics

It is becoming increasingly common for embedded devices to be fitted with a graphical display of some type. This may be a very small, simple, monochromatic LCD (in nonsmart cell phone, entry-level MP3 player, intruder alarm system, etc.). At the other end of the scale, a set-top box may have a high-resolution HDTV screen to play with. Such displays require software support, which is fully integrated into the RTOS kernel.

Since it is so common for a graphic display to be accompanied by input devices, their support is generally included in a graphics package. Possibilities include pointing devices (like a mouse), touch screens, keypads, and full keyboards.

Graphics may be used in various ways. It may be simply a means to display information—the hazard/information ("matrix") displays used on freeways is a good example. Or a display might be part of a graphic user interface (GUI) with menus, windows, icons, and the like. In each case, quite different software is required and a graphics package option, provided with an RTOS, should enable the required flexibility, without imposing a disproportionate memory footprint.

File systems

When an embedded application needs to store and process considerable quantities of data, it is clearly sensible to organize that data into some kinds of file systems. This may be in (RAM) memory, in built-in flash memory, on a flash drive, a conventional hard drive, or an optical disk (CD-ROM or DVD-ROM). Again, this capability needs fully RTOS-integrated software support. A file system must be carefully designed to handle the reentrancy requirements of a multitasking system.

Adherence to standards is particularly important with file systems. For example, using an MS-DOS compatible disk format enables the developer to leverage a well-proven design and offers complete data exchangeability with desktop systems.

4

Nucleus SE

For the remainder of this book, we will be looking in detail at how a real-time operating system (RTOS) is implemented and deployed. To do that, we will work with a specific RTOS: Nucleus SE. Even if you have no intention of using this kernel, or others related to it, understanding how it works will give you a good basis for getting to grips with any RTOS.

To understand why Nucleus SE was designed the way it is, it is useful to outline the design goals of the kernel and the objectives that I had in embarking on the project:

Simplicity—The code of the kernel should be simple, clear, well commented, and documented, and thus easy to understand. Nucleus SE is primarily intended to be useful in an educational context.

Size—It should be a small, very scalable kernel (as memory, particularly RAM, might be in short supply).

Functionality—It should have a useful level of functionality, supporting standard RTOS facilities.

8/16-bit support—It should be 8/16-bit "friendly": byte-size data should be used wherever possible; data structures should not require exotic memory addressing modes; constant data should not be copied to RAM unnecessarily.

Future—There should be a growth path from Nucleus SE to Nucleus RTOS. Users should be able to easily port code between the kernels. More importantly, their expertise should be portable. The Nucleus SE API effectively implements a subset of the Nucleus RTOS API.

Cost—The business model needed to be attractive to all potential users: 8/16-bit device developers, first-time RTOS users, and those learning about RTOS technology. It was decided to make Nucleus SE freely available and royalty-free for both commercial and educational applications; you may use and modify the code in any way that you wish.

Embedded RTOS Design. DOI: https://doi.org/10.1016/B978-0-12-822851-7.00004-7

Nucleus SE target users

The result of this approach was a kernel that may be useful by three kinds of developers:

- Programmers of 8/16-bit devices, who have a need for a simple kernel or task scheduler. It is particularly attractive if such developers are keen to acquire some RTOS usage skills or if the development is of a system where other 32-bit devices are in use, where Nucleus RTOS may be a good choice.
- Developers of embedded applications employing 32-bit devices where the software complexity does not merit the cost of a conventional, commercial RTOS. Utilizing Nucleus SE may provide some useful facilities, while offering a growth path (to Nucleus RTOS) if the application complexity increases.
- Students in education and training contexts may find Nucleus SE a useful basis for the study of RTOS technology. The skills acquired are useful later, when they commence employment in the "real world."

Design decisions and trade-offs

To achieve the aforementioned goals, several carefully considered design decisions were necessary. Details of these will be included later, when the specific features are covered, but this is a summary of the key issues.

Static configuration

Nucleus SE is a statically configured RTOS—that is, all the decisions about configuration are made at build time, not dynamically at runtime. This has multiple benefits including the simplification of data structures and a reduction on code size, as use of simplified data has this side-effect and there is no need for "create" and "delete" API calls. For most applications, dynamic object creation is really not required.

Object number

The number of objects of each type is limited in a Nucleus SE application. There may be between 1 and 16 tasks and between 0 and 16 of each other type of kernel object. This simplifies object addressing (see the following section). This limitation is not a constraint to the small applications for which the kernel is intended.

Object addressing

Objects are addressed by means of an "index," which can have values from 0 to 15. Compared with the conventional use of pointers, this can be more efficient on small processors and uses less storage—an index requires only 4 bits of storage; an address is 16–32 bits.

The scheduler

An area of the kernel's design, which was subjected to careful simplification, was the scheduler. Instead of providing a flexible mechanism with priority scheduling, round robin, and time slicing options, four separate scheduler types are available; the specific scheduler for an application is selected at configuration time.

Limited functionality

Some functionality, which is available in Nucleus RTOS, has not been implemented in Nucleus SE. In many cases, this may just be for simplicity. In others, a small loss of functionality in one area renders something else much simpler to implement. These incompatibilities are highlighted in later chapters.

Memory utilization

Since limited memory applications needed to be supported by Nucleus SE, some care was given to its memory utilization. A "classical" ROM and RAM memory architecture was assumed—ROM being used for code and constant data; RAM containing variables, stack, etc. Although a specific target may have a different scheme, the Nucleus SE code is quite flexible; `#define` symbols (`ROM` and `RAM`) are used to prefix all variables and data structures to specify their location. How this is achieved is a toolkit issue.

A key requirement was to avoid unnecessary copying of data from ROM to RAM, as RAM is likely to be in short supply. The mechanism by which this is achieved is outlined in "*Data structures*" section later in this chapter.

Application program interface implementation

The application program interface (API) for Nucleus SE is implemented in a conventional way: a C function implements each API call. These calls are grouped logically.

Although Nucleus SE API calls are not precisely the same as Nucleus RTOS, the broad functionality is emulated and mapping between the APIs is very simple. Details of the Nucleus RTOS API will be included.

Critical sections

The code for many API function calls includes sequences of instructions that manipulate kernel data. Commonly, the data may be in an invalid state during the course of these instructions, so care must be taken to avoid an interrupt occurring. Or, specifically, no code from another task or an interrupt service routine (ISR) may be allowed to run, if it might conceivably access this (currently invalid) data. Such sequences of instructions are termed critical sections.

A pair of macros are defined called `NUSE_CS_Enter()` and `NUSE_CS_Exit()`. All Nucleus SE API function code uses these to enclose a critical section; thus:

```
NUSE_CS_Enter();

<non-interruptable code>

NUSE_CS_Exit();
```

Typically, the definition of these macros will effect a disable interrupts instruction and an enable interrupts instruction, respectively. This will need to be reviewed if Nucleus SE is implemented on a different CPU architecture.

Scalability

Like all modern RTOSes, Nucleus SE is scalable. The usual technique employed, to ensure that only the used RTOS components are included, is to provide all the API functions as a library. Thus, at link time, only the referenced functions are extracted and included in the memory footprint. Nucleus RTOS uses this approach for both the kernel and all the other OS components. Nucleus SE uses a different technique.

Instead of relying on a library facility in the chosen toolkit, the Nucleus SE distribution source files all contain conditional compilation directives. To configure Nucleus SE for an application, the user needs to set a number of `#define` symbols (more on this later in this chapter). This determines which API functions are compiled and, hence, included into the application.

Nucleus SE takes this approach one step further, offering a facility that I call "extreme scalability." Several aspects of the kernel's functionality can be turned on and off or tuned in other ways using similar `#define` symbols. The user thus has a very fine-grained control over the memory usage.

Which application program interface?

Nucleus SE has its own "native" API, which will be described in detail in future chapters. For many users, just incorporating these API function calls into their code will be quite satisfactory.

Some users may prefer to use another API—either a standard or simply one with which they are familiar. The Nucleus SE API is sufficiently flexible that constructing a "wrapper" to map another API is likely to be quite straightforward.

One of the design goals of Nucleus SE was a high degree of compatibility—at the user level—with Nucleus RTOS. Although its API is different, it is designed so that the mapping is quite straightforward. A wrapper facilitating the use of the Nucleus RTOS API on Nucleus SE is available.

Services

Nucleus SE provides the range of facilities that might be expected in an RTOS.

Firstly, there is the scheduler, which is designed to be simple, but, by having four variants available, maintains flexibility. Support is available for run to completion, round robin, time slice, and priority scheduling schemes.

The Nucleus SE API includes nearly 50 service calls, which provide programmer access and control of tasks, memory partitions, signals, event flag groups, semaphores, mailboxes, queues, pipes, system time, application timers, and diagnostics.

In addition to simple task scheduling, Nucleus SE (optionally) supports task suspension. This may be "pure" (i.e., as a result of an explicit task suspend API call), it may be a "sleep" function (where the task is suspended for a specific time period) or it may be the result of another API call, where a task is blocked (i.e., conditionally suspended), pending the availability of a kernel resource. Unlike Nucleus RTOS, Nucleus SE does not support timeouts on blocking API calls.

The range of facilities enables a choice to be made from a hierarchy of intertask synchronization and communication facilities: from semaphores, through signals, event flags, mailboxes, and queues/pipes.

Parameter checking

By setting the configuration option NUSE_API_PARAMETER_CHECKING, code is included in all API functions to verify parameters—check for null pointers, valid object indexes, etc. Since this is extra code, consuming additional memory, it would normally be prudent to activate the option during debug, but turn it off for a production build.

Configuration

Nucleus SE is designed to be very configurable. This has two facets. Firstly, the kernel can be configured in a very fine-grained way to meet the needs of a specific application—tuning the available functionality and controlling memory utilization are very straightforward. Secondly, the Nucleus SE code is intended to be very portable—between toolkits and between processors.

Naming conventions

As clarity and ease of understanding of the Nucleus SE code were a goal, some thought was given to naming conventions. Every symbol in the code has the prefix NUSE_. What follows this string obeys some simple rules.

API calls

Every API call function name in Nucleus SE starts NUSE_, which is almost always followed by the type of object in question, then the operation to be performed, all in mixed case and separated by underscores; an example is NUSE_Queue_Send(), which places a message in a queue.

Other functions and variables

All other functions and (global) variables in the Nucleus SE code continue to use the NUSE_ prefix, but the remainder of the name does not necessarily have any "structure." This is unimportant to the normal user of the kernel, as the API is sole recommended interface into the code.

Configuration symbols

Since Nucleus SE is configured by means of #define symbols, these too obey the naming rules. They are also solely in uppercase. The API call enablers have exactly the same name as the functions themselves (except for the case); an example is `NUSE_QUEUE_SEND`.

Other `#define` symbols

Any other `#define` symbols—API call parameter and status return values, for example, that may be used by application code, follow the same conventions; they start with `NUSE_` and are in uppercase. An example is `NUSE_SUCCESS`.

Data structures

All RTOSes maintain a number of data structures that describe kernel objects. In most implementations, these are in the form of C structures, which are typically organized into linked lists—often doubly and may be circularly linked. This makes sense, as all the relevant data are encapsulated conveniently and list members may be added or removed, as objects are created and deleted.

In Nucleus SE, objects are all static, so arranging the object data structures into a simple list was an obvious optimization. This saved the space and complexity of forward and backward pointers. However, I decided to take the optimization one step further and not use structures at all; in Nucleus SE, all the kernel object data are represented by several simple arrays (often referred to as tables) of various types—one or more for each type of object. There were several reasons for this strategy:

- Nucleus SE was intended to be "8-bit friendly." Most small CPUs do not have an optimal means for a compiler to implement C structures. Simple arrays are much more efficient.
- Since a maximum of 16 of each object type is permitted, addressing the elements of each of these arrays requires four bits—often a byte is used. This is more efficient than an address, which are normally 16 or 32 bits.
- It is required that object data, which is constant, is kept in ROM, and not copied into RAM. Since a structure cannot (in conventional, portable C) be split between ROM and RAM, each object type might require two structures, which is overly complex. In Nucleus SE, object description tables may each be in either ROM or RAM, as required.
- Because of the great configurability of Nucleus SE ("extreme scalability"), some object description data may be optional, depending on the selected facilities. This results in wide use

of conditional compilation. A structure definition with conditional compilation directives embedded within it tends to be very hard to understand. Controlling the instantiation of individual arrays this way is quite readable.

All the object data tables adhere to the hierarchical naming convention idea mentioned earlier in this chapter. So, understanding which tables logically belong together is quite straightforward.

Key differences from Nucleus RTOS

Although Nucleus SE was designed to offer a high degree of compatibility with Nucleus RTOS, many small and larger differences remain. These will be discussed in detail in the relevant chapters, so it will suffice to summarize them here.

Object data

In Nucleus RTOS, objects are created and deleted as required. In Nucleus SE all objects are created statically and defined at build time.

Object number

An effectively indefinite number of objects of each type may be supported by Nucleus RTOS. Nucleus SE imposes a maximum of 16 instances of each object type.

Object names

Nucleus RTOS allows certain object types to be given textual names, which may be used for debug purposes. This facility is not supported by Nucleus SE.

Task blocking mechanism

The Nucleus SE API call task blocking mechanism is quite simplistic. When a resource becomes available, all waiting tasks are resumed and compete (via the scheduler) to obtain the resource. The "losers" are suspended (blocked) again. In Nucleus RTOS, the mechanism is more complex, only resuming the relevant task(s), which is more efficient.

API call timeout

When making a blocking API call, Nucleus RTOS allows the programmer to specify a timeout period after which the call will return, even if the resource has not become available. This is not supported by Nucleus SE.

Task scheduling

The Nucleus RTOS scheduler is highly efficient, flexible, and fully deterministic. Nucleus SE offers a choice of schedulers, each of which is simple and reasonably efficient with the reduced number of tasks (1−16) supported.

Task priorities

A system using Nucleus RTOS can have an effectively indefinite number of tasks, which may utilize 256 priority levels, with multiple tasks at any given priority, if required. Task priority levels may also be changed at runtime. With Nucleus SE, if the priority scheduler is selected, each task must occupy a unique priority level, which cannot be changed dynamically. There is simply one level per task, up to the maximum of 16 tasks.

Interrupt handling

Nucleus RTOS supports a sophisticated two-level ISR architecture, which allows efficient interaction between ISRs and kernel services. Nucleus SE has a simpler approach, which either allows for simple ISRs that do not interact with the kernel (unmanaged interrupts) or ISRs with a full task context save that can use API calls (managed interrupts).

Device drivers

Nucleus RTOS has a well-defined device driver architecture. Nucleus SE does not, leaving it to the user to split device management between task and ISR code.

5

The scheduler

I covered the key concepts of real-time operating system (RTOS) schedulers in Chapter 2, *Multitasking and Scheduling*. Here I will look at the facilities offered by Nucleus RTOS and, in much more detail, those provided by Nucleus SE.

Scheduling in Nucleus RTOS

Since Nucleus RTOS is a fully-fledged, well-proven, commercial RTOS, it may be assumed (correctly!) that the scheduler has been carefully designed for the requirements of such a product. It is sophisticated and flexible, giving the developer a wide range of options to address almost any conceivable real-time programming challenge.

The scheduler can support an indefinite number of tasks (only limited really by available resources) and is priority-driven. A task may be assigned a priority from 0 to 255, where 0 is the highest priority and 255 the lowest. A task's priority is dynamic—it may be changed at run time by itself or by another task in the application. Multiple tasks may be assigned the same priority level. In an extreme case, all the tasks may be assigned the same priority, which enables the implementation of a round robin (RR) or time sliced scheduling scheme.

If there are multiple tasks at the same priority, they will be scheduled in a RR fashion, in the order in which they became *Ready*. A task would need to suspend itself or relinquish control to allow the next one to be run. Tasks may also be assigned time slices (TS), which enables more controlled division of available CPU time.

Task scheduling is 100% deterministic, as might be expected from such a kernel. Tasks may also be dynamically created and destroyed, the process of which is handled by the scheduler seamlessly.

Embedded RTOS Design. DOI: https://doi.org/10.1016/B978-0-12-822851-7.00005-9

Scheduling in Nucleus SE

I designed all aspects of Nucleus SE to be broadly compatible with Nucleus RTOS, while being simpler and very memory efficient. The scheduler is no exception. It provides many of the capabilities of the Nucleus RTOS scheduler, but is somewhat restricted. Flexibility is achieved by build-time configuration.

A Nucleus SE application may have a maximum of 16 tasks (and a minimum of one, obviously). The kernel does not reserve any tasks for its own use. Although this number could theoretically be increased, the efficiency of algorithms would be compromised; a number of data structures rely on holding a task index number (0–15) in a nibble (4 bits) and these would need to be redesigned, along with their associated code.

To achieve a balance between flexibility and simplicity (and size), instead of having a single scheduler with multiple capabilities, Nucleus SE offers a choice of four types of scheduler: run-to-completion (RTC), RR, TS, and priority. The scheduler is selected statically, at configuration (build) time. This selection is described, along with more details of each scheduler type, in "Scheduler types" section.

Like all aspects of Nucleus SE, tasks are static objects. They are defined at configuration time and their priority (index) cannot be changed.

Scheduler types

As outlined earlier, Nucleus SE offers a choice of four types of scheduler. In common with most aspects of Nucleus SE configuration, this choice is determined by an entry in `nuse_config.h`—the parameter `NUSE_SCHEDULER_TYPE` needs to be set appropriately, as illustrated in this extract from the configuration file:

```
/* Scheduler type options */
#define NUSE_RUN_TO_COMPLETION_SCHEDULER     1
#define NUSE_TIME_SLICE_SCHEDULER     2
#define NUSE_ROUND_ROBIN_SCHEDULER      3
#define NUSE_PRIORITY_SCHEDULER      4
/* Scheduler type selection: */
#define NUSE_SCHEDULER_TYPE      NUSE_RUN_TO_COMPLETION_
                                                 SCHEDULER
```

Whichever scheduler is selected, its startup code is called immediately after system initialization. Full details of Nucleus SE initialization will be covered in Chapter 17, *Nucleus SE Initialization and Start-Up*.

Run-to-completion scheduler

The RTC scheduler is the simplest option and should be the first choice if it meets the demands of the application. Each task must complete its work before returning and allowing the scheduler to run the next task.

There is no requirement for a separate stack for each task and all of the code is written in C—no assembly language is needed at all. Here is the entire code of the RTC scheduler.

```
void NUSE_Scheduler(void)
{
  NUSE_TASK task_count;

  NUSE_Task_State = NUSE_TASK_CONTEXT;

  while (TRUE)
  {
    for (task_count = 0; task_count < NUSE_TASK_NUMBER;
                                        task_count++)
    {
      #if NUSE_SUSPEND_ENABLE
        if (NUSE_Task_Status[task_count] == NUSE_READY)
      #endif
        {

          #if NUSE_SCHEDULE_COUNT_SUPPORT
            NUSE_CS_Enter();
            NUSE_Task_Schedule_Count[task_count]++;
            NUSE_CS_Exit();
          #endif

          NUSE_Task_Active = task_count;
          ((PF0)NUSE_Task_Start_Address[task_count])();
        }
    }
  }
}
```

The code is really just an infinite loop, which calls each task in turn. The array NUSE_Task_Start_Address[] contains pointers to each task's outer function. The macro PF0 is simply a caste to convert the void pointer to be a pointer to a void function with no parameters. It is solely intended to aid code readability.

Conditional compilation is used to include support for optional functionality:

NUSE_SUSPEND_ENABLE determines if tasks may be suspended;

NUSE_SCHEDULE_COUNT_SUPPORT determines whether a count of each time a task is scheduled is required.

Round robin scheduler

If a little more flexibility, than that afforded by the RTC scheduler, is required, the RR scheduler may be a good choice. It provides the possibility for a task to relinquish control or suspend itself and later continue from the same point. The additional overhead, apart from code complexity and nonportability, is that each task requires its own stack.

The scheduler code is in two parts. The start-up component looks like the following:

```
void NUSE_Scheduler(void)
{
  NUSE_TASK task;

  #if NUSE_INITIAL_TASK_STATE_SUPPORT
    for (task = 0; task < NUSE_TASK_NUMBER; task++)
    {
      if (NUSE_Task_Status[task] == NUSE_READY)
      {
        break;
      }
    }
  #else
    task = 0;
  #endif

  #if NUSE_SCHEDULE_COUNT_SUPPORT
    NUSE_Task_Schedule_Count[task]++;
  #endif
  NUSE_Task_Next = task;
  NUSE_Task_State = NUSE_TASK_CONTEXT;
  NUSE_Context_Load();
}
```

If initial task state support is enabled (by a setting of NUSE_INITIAL_TASK_STATE_SUPPORT), the first *Ready* task is located; otherwise a task index of 0 is used. The context of this task is then loaded using NUSE_Context_Load(). The second part of the scheduler is the "reschedule" component:

```
void NUSE_Reschedule(void)
{
  #if NUSE_SUSPEND_ENABLE
    NUSE_Task_Next = NUSE_Task_Active;
    do
    {
      NUSE_Task_Next++;
```

```
    if (NUSE_Task_Next == NUSE_TASK_NUMBER)
        NUSE_Task_Next = 0;
  } while (NUSE_Task_Status[NUSE_Task_Next] != NUSE_READY);
#else
  NUSE_Task_Next = NUSE_Task_Active + 1;
  if (NUSE_Task_Next == NUSE_TASK_NUMBER)
  {
      NUSE_Task_Next = 0;
  }
#endif

#if NUSE_SCHEDULE_COUNT_SUPPORT
  NUSE_Task_Schedule_Count[NUSE_Task_Next]++;
#endif

/* reset time slice tick counter */
/* done here to accommodate relinquish */
/* as well as ISR count down */

#if NUSE_SCHEDULER_TYPE == NUSE_TIME_SLICE_SCHEDULER
  NUSE_Time_Slice_Ticks = NUSE_TIME_SLICE_TICKS;
#endif

  if (NUSE_Task_State != NUSE_MISR_CONTEXT)
  {
      NUSE_CONTEXT_SWAP();
  }
}
```

This code is called when a task relinquishes the CPU or is suspended.

The code selects the next task index to be run and places the value into NUSE_Task_Next, taking into account whether task suspend is enabled or not. The NUSE_CONTEXT_SWAP() macro is then used to invoke a context swap using a software interrupt.

Time slice scheduler

The TS scheduler provides just a little more functionality and control than RR, but is similar in many respects. The key difference is that, in addition to a task being able to relinquish control voluntarily, or being suspended, a task is only allowed to run for a specific number of clock ticks. This value is the same for all tasks in a Nucleus SE application that employs the TS scheduler and is set using the symbol NUSE_TIME_SLICE_TICKS in nuse_config.h.

The TS scheduler start-up code is identical to that of the RR scheduler and was outlined earlier.

The reschedule component is also identical to the RR scheduler, which was outlined earlier, but additional, conditionally compiled code resets the counter NUSE_Time_Slice_Ticks. This component is likewise called when a task relinquishes the CPU or is suspended, but is also called by the real-time clock interrupt service routine:

```
#if NUSE_SCHEDULER_TYPE == NUSE_TIME_SLICE_SCHEDULER

  if (--NUSE_Time_Slice_Ticks == 0)
  {
    NUSE_Reschedule();
  }

#endif
```

For further information on the real-time clock interrupt, see Chapter 16, *Interrupts in Nucleus SE*.

Priority scheduler

The Nucleus SE priority scheduler is, like the other options, designed to provide the required functionality, but still be quite simple. As a result, each task has a unique priority—it is not possible to have multiple tasks at any priority level. This priority is inferred from the index of the task—0 is the highest priority level. The index of a task is determined by its place in the NUSE_Task_Start_Address[] array—Chapter 20, *Using Nucleus SE*, will give more details on configuring tasks.

Like the RR and TS schedulers, the priority scheduler has two components. It shares the start-up component with the RR and TS schedulers, which was illustrated earlier. The reschedule component is slightly different:

```
void NUSE_Reschedule(U8 new_task)
{
    if (new_task == NUSE_NO_TASK)
    {
      for (new_task = 0; new_task < NUSE_TASK_NUMBER;
                                          new_task++)
      {
        if (NUSE_Task_Status[new_task] == NUSE_READY)
        {
          break;
        }
      }
    }
    else
    {
      if (new_task > NUSE_Task_Active)
```

```
    {
      return;
    }
  }

  #if NUSE_SCHEDULE_COUNT_SUPPORT
    NUSE_Task_Schedule_Count[new_task]++;
  #endif

  NUSE_Task_Next = new_task;
  if (NUSE_Task_State != NUSE_MISR_CONTEXT)
  {
    NUSE_CONTEXT_SWAP();
  }
}
```

There is no conditional code to account for task suspend not being enabled, as this capability is mandatory for the priority scheduler; any alternative would be illogical. The `NUSE_Reschedule()` function takes a parameter, which is a "hint" about which task might be scheduled next—`new_task`. This value is set when the reschedule is called because another task is woken up. That task's index is passed as the parameter. The scheduler can then determine whether to perform a context switch by comparing `new_task` with the current task's index (`NUSE_Task_Active`). If the reschedule is the result of a task being suspended, the parameter will be set to `NUSE_NO_TASK` and the scheduler simply searches for the highest priority *Ready* task.

Task states

As discussed in Chapter 2, *Multitasking and Scheduling,* all operating systems tend to have a concept of tasks being in a particular "state." The details differ from one RTOS to another. Here we will look at how Nucleus RTOS and Nucleus SE utilize task states.

Nucleus RTOS task states

Nucleus RTOS basically supports five task states:

Executing—The task which is currently in control of the CPU. Obviously only one task can occupy this state.

Ready—A task which is prepared to execute (or continue executing) pending a decision by the scheduler to run it. Typically, the task may be at a lower priority that the one that is *Executing.*

Suspended—A task which is "asleep." It is not considered for scheduling until it is woken up, at which point it will be *Ready* and may later continue executing. A task is usually asleep because it is waiting for something to occur: a resource to become available, a time period to elapse, or another task to wake it up.

Terminated—A task has been "killed." It is not considered for scheduling until it is reset, at which point it may be *Ready* or *Suspended*.

Finished—A task has completed and exited from its outer function, either by simply exiting the outer block or executing a return statement. It is not considered for scheduling until it is reset, at which point it may be *Ready* or *Suspended*.

Since Nucleus RTOS supports dynamic creation and destruction of objects—including tasks—a task might also be considered to be in the "deleted" state. However, since once a task is deleted, all of its system resources cease to exist; the task itself no longer exists, so it cannot have a state. The task's code may still be available, but the task entity would need to be created again.

Nucleus SE task states

The Nucleus SE task state model is somewhat simpler. There are normally just three states: *Executing*, *Ready* and *Suspended*. The state of each task is stored in **NUSE_Task_Status[]**, which has values like **NUSE_READY**—although it never has a value reflecting the *Executing* state. If task suspend is not enabled, only two task states are possible and this array does not exist.

There are multiple possible types of task suspend. If a task is suspended explicitly by itself or by another task, that is termed a "pure suspend" and represented by the status **NUSE_PURE_SUSPEND**. If task sleep is enabled and a task has suspended itself for a time period, it has the status **NUSE_SLEEP_SUSPEND**. If application program interface (API) function call blocking is enabled (via **NUSE_BLOCKING_ENABLE**), a task may be suspended, pending the availability of a resource. Each type of object has its own task suspend status of the form **NUSE_MAILBOX_SUSPEND**, for example. In Nucleus SE, a task may be blocked on a memory partition, an event group, a mailbox, a queue, a pipe, or a semaphore.

Thread state

The words "state" and "status" tend to be used quite loosely when discussing the behavior of tasks. There is an additional factor, which I arbitrarily term the "thread state." This is a global variable—`NUSE_Thread_State`—which carries an indication of the nature of the code being run. This is relevant to the behavior of many API calls. Possible values are:

`NUSE_TASK_CONTEXT`—The API call was made from a task.

`NUSE_STARTUP_CONTEXT`—The API call was made from the start-up code; the scheduler has not yet been started.

`NUSE_NISR_CONTEXT` and `NUSE_MISR_CONTEXT`—The API call was made from an interrupt service routine. Interrupts in Nucleus SE will be discussed in Chapter 16, *Interrupts in Nucleus SE.*

Options

I designed Nucleus SE so that as many features as possible could be optional, allowing memory and/or time to be saved, if a capability is not required. This is apparent in a number of facets of task scheduling.

Task suspend

As mentioned in "*Task states*" section, Nucleus SE supports various forms of task suspend, but this functionality is entirely optional and is enabled by the symbol `NUSE_SUSPEND_ENABLE` in `nuse_config.h`. If this symbol is set to `TRUE`, the data structure `NUSE_Task_Status[]` is defined. This has an entry for each task. The array is of type `U8` and the two nibbles are used separately. The lower nibble contains the task status: `NUSE_READY, NUSE_PURE_SUSPEND, NUSE_SLEEP_SUSPEND, NUSE_MAILBOX_SUSPEND`. If the task is suspended on an API call (e.g., `NUSE_MAILBOX_SUSPEND`), the upper nibble contains the index of the object on which it is suspended. This information is used when a resource becomes available and the API call needs to ascertain which suspended task should be woken.

A pair of scheduler functions—`NUSE_Suspend_Task()` and `NUSE_Wake_Task()`—are used to effect task suspend operations.

`NUSE_Suspend_Task()` is coded thus:

```
void NUSE_Suspend_Task(NUSE_TASK task, U8 suspend_code)
{
    NUSE_Task_Status[task] = suspend_code;
    #if NUSE_BLOCKING_ENABLE
        NUSE_Task_Blocking_Return[task] = NUSE_SUCCESS;
    #endif
    if (task == NUSE_Task_Active)
```

```
      {
        #if NUSE_SCHEDULER_TYPE == NUSE_PRIORITY_SCHEDULER
          NUSE_Reschedule(NUSE_NO_TASK);
        #else
          NUSE_Reschedule();
        #endif
      }
    }
```

This function stores the new task status (both nibbles), which it receives as a parameter: suspend_code. If blocking is enabled (see "*API call blocking*" section), the return code NUSE_SUCCESS is stored. Then NUSE_Reschedule() is called to pass control to the next qualifying task.

The code for NUSE_Wake_Task() is quite simple:

```
void NUSE_Wake_Task(NUSE_TASK task)
{
    NUSE_Task_Status[task] = NUSE_READY;
      #if NUSE_SCHEDULER_TYPE == NUSE_PRIORITY_SCHEDULER
        NUSE_Reschedule(task);
      #endif
}
```

The status of the task is set to NUSE_READY. If the priority scheduler is not in use, the current task continues to have control of the CPU until it is time to relinquish. If priority scheduling is configured, NUSE_Reschedule() is called, with the task index as a "hint," as this task may be of a higher priority and should be scheduled straight away.

API call blocking

In Nucleus RTOS, there are numerous API calls where the programmer has the option of specifying that a task be suspended (blocked) if a resource is unavailable. The task is then resumed when that resource is free again. This facility is also implemented in Nucleus SE and applies to a number of kernel objects; a task may be blocked on a memory partition, an event group, a mailbox, a queue, a pipe, or a semaphore. But, like many facilities in Nucleus SE, it is optional and determined by a symbol—NUSE_BLOCKING_ENABLE—in nuse_config.h. If this symbol is set to TRUE, the NUSE_Task_Blocking_Return[] array is defined, which carries the return code for each task; this may be NUSE_SUCCESS or codes of the form NUSE_MAILBOX_WAS_RESET that indicate an object was reset, while the task was blocked. Also, if blocking is enabled, the appropriate code is included in API functions by conditional compilation.

Schedule count

Nucleus RTOS keeps a count of how many times a task has been scheduled since it was created or last reset. This facility is also implemented in Nucleus SE, but it is optional and determined by a symbol—`NUSE_SCHEDULE_COUNT_SUPPORT`—in `nuse_config.h`. If this symbol is set to `TRUE`, the `U16` array `NUSE_Task_Schedule_Count[]`, which stores the count for each task in the application, is instantiated.

Initial task state

When a task is created in Nucleus RTOS, its state—*Ready* or *Suspended*—may be selected. In Nucleus SE, by default, all tasks are *Ready* at start-up. An option, selected using the symbol `NUSE_INITIAL_TASK_STATE_SUPPORT` in `nuse_config.h`, provides a means to select the start-up state. An array, `NUSE_Task_Initial_State[]`, is defined in `nuse_config.c` and needs to be initialized to `NUSE_READY` or `NUSE_PURE_SUSPEND` for each task in the application.

Context saving

The idea of task context saving—with any type of scheduler other than RTC—was introduced in Chapter 2, *Multitasking and Scheduling*. As I mentioned there, it is possible to save context in a number of different ways. Given that Nucleus SE is not targeted at high-end 32-bit processors, I chose not to use stack for context saving and adopted a table-based approach.

A two-dimensional array of type `ADDR`, called `NUSE_Task_Context[][]`, is used to hold contexts for all the tasks in an application. Its first dimension is `NUSE_TASK_NUMBER` (the number of tasks in the application); the second dimension is `NUSE_REGISTERS`, which is processor-dependent and sets up in `nuse_types.h`—this is the number of registers to be saved.

The context saving and restoring code is processor-dependent (obviously!) and is really the only Nucleus SE code that is so tied a specific device (and toolkit). I chose to write the example context save/restore code for the ColdFire processor. This may seem an odd choice, being a somewhat obsolete CPU, but I felt that, as the assembly language was so much easier to read than many newer devices, it would be worthwhile. The code is reasonably straightforward to follow and to use as a model to write a context switch for other processors:

```
SECTION code,4,C
ALIGN   4
XDEF    _NUSE_Context_Load
XDEF    _NUSE_Context_Swap
XREF    _NUSE_Task_Active
XREF    _NUSE_Task_Next
XREF    _NUSE_Task_Context

_NUSE_Context_Swap:

;   temporarily save D0 & A0
    move.l d0,-(sp)
    move.l a0,-(sp)

;   set up A0 to point to start of context block
    lea    _NUSE_Task_Context,a0
    clr    d0
    move.b _NUSE_Task_Active,d0
    lsl    #3,d0  ; * 8
    add.l  d0,a0
    lsl    #3,d0  ; * 64
    add.l  d0,a0  ; 8 + 64 = 72 bytes - 18 words

;   store registers
    move.l (sp)+,(a0)+      ; A0
    move.l (sp)+,(a0)+      ; D0
    movem.l d1-d7/a1-a6,(a0)    ; D1-D7 & A1-A6
    lea.l 52(a0),a0
    move.l (sp)+,(a0)+      ; SR
    move.l (sp)+,(a0)+      ; PC
    move.l sp,(a0)     ; SP

_NUSE_Context_Load:

;   set up A0 to point to end of context block
    lea    _NUSE_Task_Context,a0
    clr    d0
    move.b _NUSE_Task_Next,d0
    move.b d0,_NUSE_Task_Active
    addq   #1,d0
    lsl    #3,d0  ; * 8
    add.l  d0,a0
    lsl    #3,d0  ; * 64
    add.l  d0,a0  ; 8 + 64 = 72 bytes - 18 words
```

```
;   extract registers
    move.l -(a0),sp        ;SP
    move.l -(a0),-(sp)        ;PC
    move.l -(a0),-(sp)        ;SR
    lea.l  -52(a0),a0
    movem.l (a0),d1-d7/a1-a6    ;D1-D7 & A1-A6
    move.l -(a0),d0    ;D0
    move.l -(a0),a0    ;A0

    rte

    end
```

When a context switch is required, this code is called at the entry point NUSE_Context_Swap. Two global variables are used: NUSE_Task_Active is the index of the current task whose context is to be saved; NUSE_Task_Next is the index of the task whose context is to be loaded. See "*Global data*" section later in this chapter.

The context save process works as follows:

- Registers A0 and D0 are saved temporarily on the stack.
- A0 is set to point to the context blocks array NUSE_Task_Context [][].
- D0 is loaded with NUSE_Task_Active and multiplied by 72 (ColdFire has 18 registers requiring 72 bytes of storage).
- D0 is then added to A0, which now points to the context block for the current task.
- The registers are then saved into the context block; first A0 and D0 (from the stack), then D1-D7 and A1-A6, then the SR and PC (from the stack—we will look at how a context swap is actually initiated shortly), and lastly the stack pointer is saved.

The context load process is simply the same sequence in reverse:

- A0 is set to point to the context blocks array NUSE_Task_Context[][].
- D0 is loaded with NUSE_Task_Active, incremented and multiplied by 72.
- D0 is then added to A0, which now points to the end of the context block for the new task (as the context load must be performed in the reverse sequence to the save—the stack pointer is needed first).
- The registers are then recovered from the context block; first the stack pointer, the PC and SR are pushed onto the stack, then D1-D7 and A1-A6, and lastly D0 and A0 are loaded.

A complication, with implementing a context swap, is that access to the status register (**sr** for ColdFire) is challenging on many processors. A commonly applicable solution is to simulate an interrupt—that is, perform a software interrupt or trap—which results in the SR being pushed onto the stack along with the **pc**. This is how the ColdFire implementation of Nucleus SE works. A macro—**NUSE_CONTEXT_SWAP()**—is defined in **nuse_types.h**, which expands to:

```
asm(" trap #0");
```

This vectors through address **$80** to **NUSE_Context_Swap**.

Here is the initialization code (in NUSE_Init_Task() in nuse_init.c) for the context blocks:

```
#if NUSE_SCHEDULER_TYPE!= NUSE_RUN_TO_COMPLETION_SCHEDULER
    NUSE_Task_Context[task][15] =          /* SR */
      NUSE_STATUS_REGISTER;
    NUSE_Task_Context[task][16] =          /* PC */
      NUSE_Task_Start_Address[task];
    NUSE_Task_Context[task][17] =          /* SP */
      (U32 *)NUSE_Task_Stack_Base[task] + NUSE_Task_Stack_
                                        Size[task];
#endif
```

This only initializes the stack pointer, **pc** and the **sr**. The first two have values set by the user in **nuse_config.c**. The value of **sr** is defined as a symbol—**NUSE_STATUS_REGISTER**—in **nuse_types.h**. For ColdFire, this value is **0x40002000**.

Global data

The Nucleus SE scheduler itself requires very little data memory, but, of course, uses data structures associated with tasks, which will be covered in detail in upcoming chapters.

RAM data

There is no ROM data associated with the scheduler itself and between two and five global RAM variables (all defined in **nuse_globals.c**), depending on which scheduler is in use:

NUSE_Task_Active—this is a variable of type **u8**, which contains the index of the currently running task.

NUSE_Task_State—this is a variable of type **u8**, which contains a value that indicates the "state" of the code currently executing, which may be a task and interrupt service routine or start-up code; possible values are: NUSE_TASK_CONTEXT, NUSE_STARTUP_CONTEXT, NUSE_NISR_CONTEXT, and NUSE_MISR_CONTEXT.

NUSE_Task_Saved_State—this is a variable of type u8, which is used to preserve the value of NUSE_Task_State in managed interrupts.

NUSE_Task_Next—this is a variable of type u8, which contains the index of the next task to be scheduled for all but the RTC scheduler.

NUSE_Time_Slice_Ticks—this is a variable of type u16, which contains the TS tick counter; it is only used with the TS scheduler.

Scheduler data footprint

The Nucleus SE scheduler uses no ROM data. The exact amount of RAM data varies depending on which scheduler is in use:

For RTC, it is 2 bytes (NUSE_Task_Active and NUSE_Task_State).

For RR and priority, it is 4 bytes (NUSE_Task_Active, NUSE_Task_State, NUSE_Task_Saved_State, and NUSE_Task_Next).

For TS, it is 6 bytes (NUSE_Task_Active, NUSE_Task_State, NUSE_Task_Saved_State, NUSE_Task_Next, and NUSE_Time_Slice_Ticks).

Implementing other scheduling schemes

Although Nucleus SE is supplied with a choice of just four scheduler types, which should cover most eventualities, the open architecture lends itself to the implementation of some other possibilities.

Time slice with background

As discussed in Chapter 2, *Multitasking and Scheduling*, a simple TS scheduler has limitations, as it simply bounds the maximum time that a task can "hog" the CPU. A more sophisticated option would be to add support for a background task. This task would be scheduled in any slots allocated to tasks that are suspended and would also run if a task relinquished the remainder of a slot. This approach results in tasks being scheduled at strictly regular intervals and they each receive a fully predictable proportion of CPU time.

Priority and round robin

In many real-time kernels, the priority scheduler supports multiple tasks at each priority level, unlike Nucleus SE where each task has a unique level. I made this choice because it greatly simplified the data structures and, hence, the scheduler code. Multiple tables in both ROM and RAM would be needed to support the more sophisticated architecture.

6

Tasks

The concept of a task was introduced in earlier chapters. In most respects, Nucleus SE conforms to expectations in terms of how a task is represented at the conceptual level. Fundamentally, a task is just a set of register values, which may be currently loaded into the processor's registers (for the executing task) or may be stored ready for a context switch to a task at some later time. In most cases, a task also has a stack space associated with it.

With a run-to-completion (RTC) scheduler of course, there is no context switch and a task can be considered to be just a program counter (PC) (code entry point) value.

The definition of a task does not encompass the actual code. Of course, a task needs to execute code, but does not "own" it. Functions may be shared between multiple tasks. Multiple tasks may even share *all* their code. Shared code must almost always be written to be reentrant—that is, not use static data, which is not sharable. Most compilers accommodate this with little problem, but care may be required with library functions, as they may not have been designed for a multitasking application.

This definition drives the design of task data structures and application program interface (API) functions that are described in this chapter. I am going to take a look at how tasks are configured in Nucleus SE and detail the service calls (API) that appertain to tasks in both Nucleus SE and Nucleus RTOS.

Configuring tasks

Number of tasks

As with most aspects of Nucleus SE, the configuration of tasks is primarily controlled by #define statements in nuse_config.h. The key setting is NUSE_TASK_NUMBER, which determines how many tasks are configured for the application. The default setting is 1 (i.e., a single task in use) and you can set it to any value up to 16. An erroneous value will result in a compile time error, which is generated by a test in nuse_config_check.h (this is included into

Embedded RTOS Design. DOI: https://doi.org/10.1016/B978-0-12-822851-7.00006-0

`nuse_config.c` and hence compiled with this module) resulting in a `#error` statement being compiled. This setting results in some data structures being defined and sized accordingly, of which more shortly.

In Nucleus SE, except with the RTC scheduler, it is essential that at least one task is always ready to run. With the Priority scheduler, you should simply ensure that the lowest priority task is never suspended; think of it as a "background task."

Unlike some other real-time kernels, Nucleus SE does not utilize any "system tasks," so all 16 tasks are available for use by user application code or middleware.

API enables

Every API function (service call) in Nucleus SE has an enabling `#define` symbol in `nuse_config.h`. For tasks, these are:

```
NUSE_TASK_SUSPEND
NUSE_TASK_RESUME
NUSE_TASK_SLEEP
NUSE_TASK_RELINQUISH
NUSE_TASK_CURRENT
NUSE_TASK_CHECK_STACK
NUSE_TASK_RESET
NUSE_TASK_INFORMATION
NUSE_TASK_COUNT
```

By default, all of these are set to `FALSE`, thus disabling each service call and inhibiting the inclusion of any implementation code. To configure tasks for an application, you need to select the API calls that you want to use and set their enabling symbols to `TRUE`.

Here is an extract from the default `nuse_config.h` file.

```
/*** Tasks and task control ***/

#define NUSE_TASK_NUMBER 1     /* Number of tasks in the
                                  system - 1-16 */

#define NUSE_TASK_SUSPEND  FALSE      /* Service call enabler */
#define NUSE_TASK_RESUME  FALSE       /* Service call enabler */
#define NUSE_TASK_SLEEP  FALSE        /* Service call enabler */
#define NUSE_TASK_RELINQUISH  FALSE   /* Service call enabler */
#define NUSE_TASK_CURRENT  FALSE      /* Service call enabler */
#define NUSE_TASK_CHECK_STACK  FALSE  /* Service call enabler */
#define NUSE_TASK_RESET  FALSE        /* Service call enabler */
#define NUSE_TASK_INFORMATION  FALSE  /* Service call enabler */
#define NUSE_TASK_COUNT  FALSE        /* Service call enabler */
```

If your code uses an API call, which has not been enabled, a link time error will result, as no implementation code will have been included in the application.

Functionality enables

In Nucleus SE, a number of aspects of tasks' functionality may also be optionally enabled. Again, symbols in the `nuse_config.h` file are utilized:

NUSE_SUSPEND_ENABLE—This enables tasks to be placed into a *Suspended* state. If this option is not set, all tasks are always ready to be scheduled. Clearly, setting this option is mandatory when the Priority scheduler is in use.

NUSE_BLOCKING_ENABLE—This enables tasks to be suspended on a number of API function calls. If this option is selected, NUSE_SUSPEND_ENABLE is also required.

NUSE_INITIAL_TASK_STATE_SUPPORT—This enables the start-up state of a task to be specified. If this option is not selected, all tasks start-up ready to be scheduled.

Task service calls

Nucleus RTOS supports 16 service calls (APIs) that appertain to tasks, which provide the following functionality:

Suspend a task: NU_Suspend_Task(). Implemented by NUSE_Task_Suspend() in Nucleus SE.

Wake up a task (resume): NU_Resume_Task(). Implemented by NUSE_Task_Resume() in Nucleus SE.

Put a task to sleep for a specified period: NU_Sleep(). Implemented by NUSE_Task_Sleep() in Nucleus SE.

Relinquish control of the processor: NU_Relinquish(). Implemented by NUSE_Task_Reliquish() in Nucleus SE.

Obtain the current task's ID: NU_Current_Task_Pointer(). Implemented by NUSE_Task_Current() in Nucleus SE.

Check available stack space: NU_Check_Stack(). Implemented by NUSE_Task_Check_Stack() in Nucleus SE.

Restore a task to the unused state (reset): NU_Reset_Task(). Implemented by NUSE_Task_Reset() in Nucleus SE.

Provide information about a specified task: `NU_Task_Information()`. Implemented by `NUSE_Task_Information()` in Nucleus SE.

Return a count of how many tasks are (currently) configured for the application: `NU_Established_Tasks()`. Implemented by `NUSE_Task_Count()` in Nucleus SE.

Add a new task to the application (create): `NU_Create_Task()`. Not implemented in Nucleus SE.

Remove a task from the application (delete): `NU_Delete_Task()`. Not implemented in Nucleus SE.

Return pointers to all the tasks (currently) in the application: `NU_Task_Pointers()`. Not implemented in Nucleus SE.

Change a task's preemption posture: `NU_Change_Preemption()`. Not implemented in Nucleus SE.

Change a task's priority: `NU_Change_Priority()`. Not implemented in Nucleus SE.

Change a task's time slice: `NU_Change_Time_Slice()`. Not implemented in Nucleus SE.

Terminate a task: `NU_Terminate_Task()`. Not implemented in Nucleus SE.

The implementation of each of these service calls is examined in detail through the rest of this chapter.

Task control services

The fundamental controls that may be applied to a task are suspending indefinitely, resuming, suspending for a fixed time period (sleep), and relinquishing control of the CPU. Nucleus RTOS and Nucleus SE each provide four basic API calls for these operations, which I will discuss here.

Suspending a task

Nucleus RTOS provides a simple API call to enable a specified task to be suspended indefinitely. Nucleus SE has a call with the same functionality.

Nucleus RTOS API call for task suspend
Service call prototype:

```
STATUS NU_Suspend_Task(NU_TASK *task);
```

Parameters:

> **task**—pointer to the task control block of the task to be suspended (which may be the current task, the ID of which may be obtained using **NU_Current_Task_Pointer()**)

Returns:

> **NU_SUCCESS**—The call was completed successfully.

> **NU_INVALID_TASK**—The task pointer is invalid.

> **NU_INVALID_SUSPEND**—The specified task has a status of **NU_FINISHED** or **NU_TERMINATED**

Nucleus SE API call for task suspend

This API call supports the key functionality of the Nucleus RTOS API.

Service call prototype:

> **STATUS NUSE_Task_Suspend(NUSE_TASK task);**

Parameters:

> **task**—the index (ID) of the task to be suspended (which may be the current task, the ID of which may be obtained using **NUSE_Task_Current()**)

Returns:

> **NUSE_SUCCESS**—The call was completed successfully.
> **NUSE_INVALID_TASK**—The task index is invalid.

Nucleus SE implementation of task suspend

The key functionality of the API function is quite simple:

```
STATUS NUSE_Task_Suspend(NUSE_TASK task)
{
  #if NUSE_API_PARAMETER_CHECKING
    if (task >= NUSE_TASK_NUMBER)
    {
        return NUSE_INVALID_TASK;
    }
  #endif

  NUSE_CS_Enter();

  NUSE_Suspend_Task(task, NUSE_PURE_SUSPEND);

  NUSE_CS_Exit();

  return NUSE_SUCCESS;
}
```

This essentially calls the scheduler utility function **NUSE_Suspend_Task()**, specifying an unconditional suspend (**NUSE_PURE_SUSPEND**). This function will invoke the scheduler if the task being suspended is current.

Resuming a task

Nucleus RTOS provides a simple API call to enable a specified task, which was previously suspended indefinitely, to be resumed. Nucleus SE has a call with the same functionality.

Nucleus RTOS API call for task resume

Service call prototype:

    STATUS NU_Resume_Task(NU_TASK *task);

Parameters:

task—pointer to the task control block of the task to be resumed

Returns:

NU_SUCCESS—The call was completed successfully.

NU_INVALID_TASK—The task pointer is invalid.

NU_INVALID_RESUME—The task was not unconditionally suspended.

Nucleus SE API call for task resume

This API call supports the key functionality of the Nucleus RTOS API.

Service call prototype:

    STATUS NUSE_Task_Resume(NUSE_TASK task);

Parameters:

task—the index (ID) of the task to be resumed

Returns:

NUSE_SUCCESS—The call was completed successfully.

NUSE_INVALID_TASK—The task index is invalid.

NUSE_INVALID_RESUME—The task was not unconditionally suspended.

Nucleus SE implementation of task resume

The key functionality of the API function is quite simple:

```
STATUS NUSE_Task_Resume(NUSE_TASK task)
{
  #if NUSE_API_PARAMETER_CHECKING
    if (task >= NUSE_TASK_NUMBER)
    {
      return NUSE_INVALID_TASK;
    }
    if (NUSE_Task_Status[task] != NUSE_PURE_SUSPEND)
    {
      return NUSE_INVALID_RESUME;
    }
  #endif

  NUSE_CS_Enter();

  #if NUSE_TIMEOUT_SUPPORT
    NUSE_Task_Timeout_Counter[task] = 0;
  #endif
  NUSE_Wake_Task(task);

  NUSE_CS_Exit();

  return NUSE_SUCCESS;
}
```

This essentially calls the scheduler utility function `NUSE_Wake_Task()`. This function will invoke the scheduler, if the Priority scheduler is in use, in case the resumed task is higher priority than the current task.

Putting a task to sleep

Nucleus RTOS provides a simple API call to enable the current task to be suspended for a specific time period. Nucleus SE has a call with the same functionality.

Nucleus RTOS API call for task sleep

Service call prototype:

```
VOID NU_Sleep(UNSIGNED ticks);
```

Parameters:

`ticks`—the time period, in real-time clock ticks, for which the task should be suspended

Returns:

Nothing

Nucleus SE API call for task sleep

This API call supports the key functionality of the Nucleus RTOS API.

Service call prototype:

```
void NUSE_Task_Sleep(U16 ticks);
```

Parameters:

ticks—the time period, in real-time clock ticks, for which the task should be suspended

Returns:

Nothing

Nucleus SE implementation of task sleep

The key functionality of the API function is quite simple:

```
void NUSE_Task_Sleep(U16 ticks)
{
  NUSE_CS_Enter();

  NUSE_Task_Timeout_Counter[NUSE_Task_Active] = ticks;
  NUSE_Suspend_Task(NUSE_Task_Active, NUSE_SLEEP_SUSPEND);

  NUSE_CS_Exit();
}
```

The code loads the delay value into the current task's entry in NUSE_Task_Timeout_Counter[]. Then the task is suspended, using a call to NUSE_Suspend_Task() specifying sleep suspend (NUSE_SLEEP_SUSPEND).

The timeout value is utilized by the real-time clock interrupt service routine. The code is shown here and is discussed in more detail in Chapter 16, *Interrupts in Nucleus SE*.

```
U8 task;

for (task = 0; task < NUSE_TASK_NUMBER; task++)
{
  if (NUSE_Task_Timeout_Counter[task] != 0)
  {
    NUSE_Task_Timeout_Counter[task]--;
    if (NUSE_Task_Timeout_Counter[task] == 0)
```

```
    {
        NUSE_Wake_Task(task);
    }
  }
}
```

Relinquishing the CPU

Nucleus RTOS provides a simple API call to enable a task to relinquish the CPU to any other *Ready* tasks, at the same priority level, on a round robin basis. Nucleus SE has a call with very similar functionality. However, it may not be used with the Priority scheduler, as only one task at each priority level is permitted. An error will result if you attempt to enable this API call with the Priority scheduler. The call works with the Round Robin and Time Slice schedulers and has no effect with the RTC scheduler.

Nucleus RTOS API call for task relinquish

This API call supports the key functionality of the Nucleus RTOS API.

Service call prototype:

```
VOID NU_Relinquish(VOID);
```

Parameters:

None

Returns:

Nothing

Nucleus SE API call for task relinquish

This API call supports the key functionality of the Nucleus RTOS API.

Service call prototype:

```
void NUSE_Task_Relinquish(void);
```

Parameters:

None

Returns:

Nothing

Nucleus SE implementation of task relinquish

The key functionality of the API function is quite simple:

```
void NUSE_Task_Relinquish(void)
{
  NUSE_CS_Enter();
  NUSE_Reschedule();   /* next Ready task in round robin will
                                              be scheduled */
  NUSE_CS_Exit();
}
```

This essentially calls the scheduler function `NUSE_Reschedule()`. This function simply invokes the scheduler to run the next task.

Task utility services

Some additional API calls, that appertain to tasks, include obtaining a task's ID, checking stack space, resetting a task, obtaining information about a task, and ascertaining the number of tasks in a system. Nucleus RTOS and Nucleus SE each provide four basic API calls for these operations, which I will discuss here.

Obtaining the current task's ID

This service call simply returns the ID of the calling task. For Nucleus RTOS, this is a pointer to the currently active task control block. For Nucleus SE, it is the index (0–15) of the current task.

Nucleus RTOS API call for current task

Service call prototype:

```
NU_TASK *NU_Current_Task_Pointer(VOID);
```

Parameters:

None

Returns:

Pointer to the currently active task control block.

`NU_NULL`—There is no currently active task.

Nucleus SE API call for current task

This API call supports the key functionality of the Nucleus RTOS API.

Service call prototype:

```
NUSE_TASK NUSE_Task_Current(void);
```

Parameters:

None

Returns:

Index of the current (calling) task

Nucleus SE implementation of current task

The implementation of this API call is almost trivially simple: the value of the global variable NUSE_Task_Active is returned.

Checking available stack space

This service call returns the available stack space (in bytes) for the current task. This only makes sense for schedulers where each task has its own stack; that is, not RTC with Nucleus SE.

Nucleus RTOS API call for stack space

Service call prototype:

```
UNSIGNED NU_Check_Stack(VOID);
```

Parameters:

None

Returns:

Number of bytes of available stack for the current task

Nucleus SE API call for stack space

This API call supports the basic functionality of the Nucleus RTOS API. However, in Nucleus SE, a dummy parameter is required in order to easily obtain the current stack pointer (SP) value (without resort to assembly code).

Service call prototype:

```
U16 NUSE_Task_Check_Stack(U8 dummy);
```

Parameters:

~~dummy~~—Any value may be provided, as it is not actually used.

Returns:

Number of bytes of available stack for the current task

Nucleus SE implementation of stack space

The code for this call is designed to be portable:

```
U16 NUSE_Task_Check_Stack(U8 dummy)
{
  #if NUSE_SCHEDULER_TYPE == NUSE_RUN_TO_COMPLETION_
                                              SCHEDULER
    return 0;
  #else
    U8 *sp;

    sp = &dummy - 1;
    return (((U8 *)NUSE_Task_Stack_Base[NUSE_Task_Active] - sp) *
                                        sizeof(unsigned));
  #endif
}
```

If the RTC scheduler is in use, a value of 0 is returned, as it is not possible (using code that is remotely portable) to determine the available stack space.

Otherwise, the value of the SP is determined by finding the address of the dummy parameter, which would be near the top of the stack. This approach is, strictly speaking, toolkit/compiler dependent, but will usually be valid. The returned value is the difference between this value and the initial stack value, converted to bytes.

Resetting a task

This API call restores the task to its initial, unused state. This API function is unusual, compared with "reset" API functions for other kernel objects, as, although it is a reset, it does not simply initialize the task to its start-up state (with Nucleus SE this is either NUSE_READY or an entry from NUSE_Task_Initial_State[]—see "*Data structures*" section); the task is placed into a *Suspended* state (NUSE_PURE_SUSPEND) and needs to be resumed before it will get scheduled. This behavior is compatible with the functionality of the equivalent Nucleus RTOS API call.

Nucleus RTOS API call for task reset

Service call prototype:

```
STATUS NU_Reset_Task(
NU_TASK *task,
UNSIGNED argc,
VOID *argv);
```

Parameters:

task—pointer to the task control block,

argc—data element that may be used to pass information to the task, and

argv—pointer that may be used to pass information to the task.

Returns:

NU_SUCCESS—The call was completed successfully.

NU_INVALID_TASK—The task pointer is not valid.

NU_NOT_TERMINATED—The specified task is not in a *Terminated* or *Finished* state; only tasks in a *Terminated* or *Finished* state can be reset.

Nucleus SE API call for task reset

This API call supports the key functionality of the Nucleus RTOS API.

Service call prototype:

```
STATUS NUSE_Task_Reset(NUSE_TASK task);
```

Parameters:

task—the index (ID) of the task to be reset

Returns:

NUSE_SUCCESS—The call was completed successfully.

NUSE_INVALID_TASK—The task index is not valid.

Nucleus SE implementation of task reset

The main job of the NUSE_Task_Reset() API function—after parameter checking—is to simply reinitialize all the task's data structures:

```
STATUS NUSE_Task_Reset(NUSE_TASK task)
{
  #if NUSE_API_PARAMETER_CHECKING
    if (task >= NUSE_TASK_NUMBER)
    {
      return NUSE_INVALID_TASK;
    }
  #endif

  NUSE_CS_Enter();

  #if NUSE_BLOCKING_ENABLE      /* if task was blocked on API call */
    switch (LONIB(NUSE_Task_Status[task]))   /* need to update
                                                kernel object */
                          /* blocking info */
    {
      case NUSE_MAILBOX_SUSPEND:
        NUSE_Mailbox_Blocking_Count[HINIB(NUSE_Task_Status
                                              [task])]--;
        break;
      case NUSE_SEMAPHORE_SUSPEND:
        NUSE_Semaphore_Blocking_Count[HINIB(NUSE_Task_
                                          Status[task])]--;
        break;
      case NUSE_PARTITION_SUSPEND:
        NUSE_Partition_Pool_Blocking_Count[
          HINIB(NUSE_Task_Status[task])]--;
        break;
      case NUSE_QUEUE_SUSPEND:
        NUSE_Queue_Blocking_Count[HINIB(NUSE_Task_Status
                                              [task])]--;
        break;
      case NUSE_PIPE_SUSPEND:
        NUSE_Pipe_Blocking_Count[HINIB(NUSE_Task_Status
                                              [task])]--;
        break;
      case NUSE_EVENT_SUSPEND:
        NUSE_Event_Group_Blocking_Count[HINIB(NUSE_Task_
                                          Status[task])]--;
        break;
    }
  #endif

  NUSE_Init_Task(task);
  NUSE_Suspend_Task(task, NUSE_PURE_SUSPEND);

  NUSE_CS_Exit();

  return NUSE_SUCCESS;
}
```

If the task is blocked on an API call, awaiting access to a kernel object, the first requirement is to adjust the blocked task counter associated with that object. This is effected by means of an initial switch statement.

Then, the task's data structures are initialized (mainly to zeros, except for its context block) by calling the initialization routine `NUSE_Init_Task()`. The operation of this function is covered in more detail in Chapter 17, *Nucleus SE Initialization and Start-up*, where we look at system initialization. Lastly, the task's status is set to `NUSE_PURE_SUSPEND`.

Obtaining task information

This service call obtains a selection of information about a task. The Nucleus SE implementation differs from Nucleus RTOS in that it returns less information, as object naming, preemption option, and time slice are not supported and priority would be redundant.

Nucleus RTOS API call for obtain task information

Service call prototype:

```
STATUS NU_Task_Information(
NU_TASK *task,
CHAR *name,
DATA_ELEMENT *task_status,
UNSIGNED *scheduled_count,
OPTION *priority,
OPTION *preempt,
UNSIGNED *time_slice,
VOID **stack_base,
UNSIGNED *stack_size,
UNSIGNED *minimum_stack;
```

Parameters:

`task`—pointer to the task about which information is being requested;

`name`—pointer to an 8-character destination area for the task's name; this includes space for the `NULL` terminator;

`task_status`—a pointer to a variable, which will receive the current value of the task status;

`scheduled_count`—a pointer to a variable which will receive a count of the number times this task has been scheduled;

`priority`—pointer to a variable to hold the task's priority;

preempt—pointer to a variable to hold the task's preempt option; **NU_PREEMPT** indicates the task is preemptable, while **NU_NO_PREEMPT** indicates the task is not preemptable;

time_slice—pointer to a variable to hold the task's time slice value; a value of zero indicates that time slicing for this task is disabled;

stack_base—a pointer to a variable which will receive the address of the task's stack;

stack_size—a pointer to a variable which will receive the size of the tasks;

minimum_stack—pointer to a variable to hold the minimum amount of bytes left in the task's stack.

Returns:

NU_SUCCESS—The call was completed successfully.

NU_INVALID_TASK—The task pointer is not valid.

Nucleus SE API call for obtain task information

This API call supports the key functionality of the Nucleus RTOS API.

Service call prototype:

```
STATUS NUSE_Task_Information(
NUSE_TASK task,
U8 *task_status,
U16 *scheduled_count,
ADDR *stack_base,
U16 *stack_size);
```

Parameters:

task—the index of the task about which information is being requested;

task_status—a pointer to a **U8** variable, which will receive the current value of the task status (nothing returned if task suspend is disabled);

scheduled_count—a pointer to a **U16** variable, which will receive a count of the number times this task has been scheduled (nothing returned if schedule count is disabled);

stack_base—a pointer to an **ADDR** variable, which will receive the address of the task's stack (nothing returned if the RTC scheduler is in use);

stack_size—a pointer to a U16 variable, which will receive the size of the task's stack (nothing returned if the RTC scheduler is in use).

Returns:

NUSE_SUCCESS—The call was completed successfully.

NUSE_INVALID_TASK—The task index is not valid.

NUSE_INVALID_POINTER—One or more of the pointer parameters is invalid.

Nucleus SE implementation of obtain task information

The implementation of this API call is quite straightforward:

```
STATUS NUSE_Task_Information(NUSE_TASK task, U8 *task_status,
    U16 *scheduled_count, ADDR *stack_base, U16 *stack_size)
{
    #if NUSE_API_PARAMETER_CHECKING
        if (task >= NUSE_TASK_NUMBER)
        {
            return NUSE_INVALID_TASK;
        }

        if ((task_status == NULL) || (scheduled_count == NULL)
          || (stack_base == NULL) || (stack_size == NULL))
        {
            return NUSE_INVALID_POINTER;
        }
    #endif

    NUSE_CS_Enter();
    #if NUSE_SUSPEND_ENABLE || NUSE_INCLUDE_EVERYTHING
        *task_status = LONIB(NUSE_Task_Status[task]);
    #endif

    #if NUSE_SCHEDULE_COUNT_SUPPORT || NUSE_INCLUDE_EVERYTHING
        *scheduled_count = NUSE_Task_Schedule_Count[task];
    #endif

    #if NUSE_SCHEDULER_TYPE! =
        NUSE_RUN_TO_COMPLETION_SCHEDULER
        *stack_base = NUSE_Task_Stack_Base[task];
        *stack_size = NUSE_Task_Stack_Size[task];
    #endif

    NUSE_CS_Exit();
    return NUSE_SUCCESS;
}
```

The function returns the task status, taking into account the various configuration options.

Obtaining the number of tasks

This service call returns the number of tasks configured in the application. Whilst in Nucleus RTOS this will vary over time and the returned value will represent the *current* number of tasks, in Nucleus SE the value returned is set at build time and cannot change.

Nucleus RTOS API call for number of tasks

Service call prototype:

```
UNSIGNED NU_Established_Tasks(VOID);
```

Parameters:

None

Returns:

The number of established (i.e., created, but not deleted) tasks in the application

Nucleus SE API call for number of tasks

This API call supports the key functionality of the Nucleus RTOS API.

Service call prototype:

```
U8 NUSE_Task_Count(void);
```

Parameters:

None

Returns:

The number of configured tasks in the application

Nucleus SE implementation of number of tasks

The implementation of this API call is almost trivially simple: the value of the **#define** symbol **NUSE_TASK_NUMBER** is returned.

Data structures

Tasks utilize a number of data structures—in RAM and ROM—which, as is the case with other Nucleus SE objects, are a series of tables, included and dimensioned according to the number of tasks configured and options selected.

I strongly recommend that application code does not access these data structures directly, but uses the provided API functions. This avoids incompatibility with possible future versions of Nucleus SE and unwanted side effects and simplifies porting of an application to Nucleus RTOS. The details of data structures are included here to facilitate easier understanding of the working of the service call code and for debugging.

Kernel RAM data

These data structures are:

NUSE_Task_Context[][]—This is a two-dimensional array of type **ADDR**, with one row for each configured task. The number of columns is chip-specific and determined by the symbol **NUSE_REGISTERS**, which is defined in **nuse_types.h**. This array is used by the scheduler to save the context for each task and was described in detail under "Context saving" section in Chapter 5, *The Scheduler*. It does not exist if the RTC scheduler is in use.

NUSE_Task_Signal_Flags[]—This array of type **U8** is created, if signals are enabled and contains the eight signal flags for each task. Signals will be discussed in Chapter 8, *Signals*.

NUSE_Task_Timeout_Counter[]—If the **NUSE_Task_Sleep()** API call is enabled, this **U16** array is created to contain the down counters for each task.

NUSE_Task_Status[]—This **U8** array contains the status of each task—**NUSE_READY** or a *Suspended* status. It is only created is task suspend is enabled.

NUSE_Task_Blocking_Return[]—If API call blocking is enabled, this **U8** array is created. It carries the return code, which will be used after an API call has been blocked. Normally it will contain **NUSE_SUCCESS** or a code to indicate the object was reset (e.g., **NUSE_MAILBOX_WAS_RESET**).

NUSE_Task_Schedule_Count[]—This **U16** array only exists if schedule counting has been enabled and contains the counts for each task.

`NUSE_Task_Context[][]` is initialized mostly to zeros, except for the entries for status register (SR), PC, and SP, which are set to initial values (see "*ROM data*" section) and all the other data structures are set to zeros by `NUSE_Init_Task()` when Nucleus SE starts up. A full description of Nucleus SE start-up procedures can be found in Chapter 17, *Nucleus SE Initialization and Start-up*.

Here are the definitions of these data structures in `nuse_init.c` file.

```
#if NUSE_SCHEDULER_TYPE! = NUSE_RUN_TO_COMPLETION_SCHEDULER
    RAM ADDR NUSE_Task_Context[NUSE_TASK_NUMBER][NUSE_REGISTERS];
#endif

#if NUSE_SIGNAL_SUPPORT || NUSE_INCLUDE_EVERYTHING
    RAM U8 NUSE_Task_Signal_Flags[NUSE_TASK_NUMBER];
#endif

#if NUSE_TASK_SLEEP || NUSE_INCLUDE_EVERYTHING
    RAM U16 NUSE_Task_Timeout_Counter[NUSE_TASK_NUMBER];
#endif

#if NUSE_SUSPEND_ENABLE || NUSE_INCLUDE_EVERYTHING
    RAM U8 NUSE_Task_Status[NUSE_TASK_NUMBER];
#endif

#if NUSE_BLOCKING_ENABLE || NUSE_INCLUDE_EVERYTHING
    RAM U8 NUSE_Task_Blocking_Return[NUSE_TASK_NUMBER];
#endif

#if NUSE_SCHEDULE_COUNT_SUPPORT || NUSE_INCLUDE_EVERYTHING
    RAM U16 NUSE_Task_Schedule_Count[NUSE_TASK_NUMBER];
#endif
```

User RAM data

It is the user's responsibility to define a stack for each task (unless the RTC scheduler is in use). These should be arrays of type `ADDR` and are normally defined in `nuse_config.c`. The addresses and sizes of the stacks need to be placed the tasks' entries in `NUSE_Task_Stack_Base[]` and `NUSE_Task_Stack_Size[]`, respectively (see "*ROM data*" section).

ROM data

Between one and four data structures appertaining to tasks are stored in ROM. The exact number depends on which options have been selected:

NUSE_Task_Start_Address[]—This is an array of type **ADDR**, with one entry for each configured task; this is a pointer to the entry point of the code for the task.

NUSE_Task_Stack_Base[]—This is an array of type **ADDR**, with one entry for each configured task; this is a pointer to the base address of the stack for the task. This array only exists if a scheduler other than RTC is selected.

NUSE_Task_Stack_Size[]—This is an array of type **U16**, with one entry for each configured task; this is the size of the stack (in words) for the task. This array only exists if a scheduler other than RTC is selected.

NUSE_Task_Initial_State[]—This is an array of type **U8**, with one entry for each configured task; this is the initial (start-up) state for the task. It may have values of **NUSE_READY** or **NUSE_PURE_SUSPEND**. This array only exists if initial task state support is selected.

These data structures are declared and initialized (statically, of course) in **nuse_config.c**; thus:

```
ROM ADDR NUSE_Task_Start_Address[NUSE_TASK_NUMBER] =
{
  /* addresses of task entry functions ------ */
  NUSE_Idle_Task
};

#if NUSE_SCHEDULER_TYPE! = NUSE_RUN_TO_COMPLETION_SCHEDULER

  /* define stack storage - arrays of type ADDR - here */
  ROM ADDR NUSE_Task_Stack_Base[NUSE_TASK_NUMBER] =
  {
    /* addresses of task stacks ------ */
  };

  ROM U16 NUSE_Task_Stack_Size[NUSE_TASK_NUMBER] =
  {
    /* stack sizes ------ */
  };

#endif

#if NUSE_INITIAL_TASK_STATE_SUPPORT || NUSE_INCLUDE_
                                              EVERYTHING
  ROM U8 NUSE_Task_Initial_State[NUSE_TASK_NUMBER] =
  {
    /* task states ------ */
    /* may be NUSE_READY or NUSE_PURE_SUSPEND */
  };

#endif
```

Task data footprint

Like all kernel objects in Nucleus SE, the amount of data memory required for tasks is readily predictable.

The ROM data footprint (in bytes) for all the tasks in an application is:

`NUSE_TASK_NUMBER * sizeof(ADDR)`

Plus, if the scheduler is not RTC:

`NUSE_TASK_NUMBER * (sizeof(ADDR) + 2)`

Plus, if initial task state support is selected:

`NUSE_TASK_NUMBER`

The RAM data footprint (in bytes) for all the tasks in an application is totally governed by selected options and may be zero if none of them are selected.

If the scheduler is not RTC:

`NUSE_TASK_NUMBER * NUSE_REGISTERS * sizeof(ADDR)`

Plus, if support for signals is selected:

`NUSE_TASK_NUMBER`

Plus, if the `NUSE_Task_Sleep()` API call is enabled:

`NUSE_TASK_NUMBER * 2`

Plus, if task suspend is enabled:

`NUSE_TASK_NUMBER`

Plus, if API call blocking is enabled:

`NUSE_TASK_NUMBER`

Plus, if schedule counting is enabled:

`NUSE_TASK_NUMBER * 2`

Unimplemented API calls

Seven task API calls found in Nucleus RTOS are not implemented in Nucleus SE:

Create task

This API call creates an application task. It is not needed with Nucleus SE, as tasks are created statically.
Service call prototype:

```
STATUS NU_Create_Task(
NU_TASK *task,
CHAR *name,
VOID (*task_entry)(UNSIGNED, VOID *),
UNSIGNED argc,
VOID *argv,
VOID *stack_address,
UNSIGNED stack_size,
OPTION priority,
UNSIGNED time_slice,
OPTION preempt,
OPTION auto_start);
```

Parameters:

task—pointer to a user-supplied task control block; this will be used as a "handle" for the task in other API calls.

name—pointers to a 7-character, NULL-terminated name for the task.

task_entry—specifies the entry function for the task.

argc—an UNSIGNED data element that may be used to pass initial information to the task.

argv—a pointer that may be used to pass information to the task.

stack_address—designates the starting memory location of the task's stack.

stack_size—specifies the number of bytes in the stack.

priority—specifies the task's priority value; between 0 and 255 with low numbers indicating highest priority.

time_slice—indicates the maximum number of timer ticks that can expire while executing this task; 0 disables time slicing for this task.

preempt—indicates whether the task is preemptable; valid values are NU_PREEMPT and NU_NO_PREEMPT.

auto_start—indicates the initial state of the task; NU_START makes is *Ready*; NU_NO_START makes it *Suspended*.

Returns:

NU_SUCCESS—indicates successful completion of the service.

NU_INVALID_TASK—indicates the task control block pointer is **NULL**.

NU_INVALID_ENTRY—indicates the task's entry function pointer is **NULL**.

NU_INVALID_MEMORY—indicates the memory area specified by **stack_address** is **NULL**.

NU_INVALID_SIZE—indicates the specified stack size is not large enough.

NU_INVALID_PREEMPT—indicates that the **preempt** parameter is invalid.

NU_INVALID_START—indicates the **auto_start** parameter is invalid.

Delete task

This API call deletes a previously created application task, which must be *Finished* or *Terminated*. It is not needed with Nucleus SE, as tasks are created statically and cannot be deleted.

Service call prototype:

```
STATUS NU_Delete_Task(NU_TASK *task);
```

Parameters:

task—pointer to task control block

Returns:

NU_SUCCESS—indicates successful completion of the service.

NU_INVALID_TASK—indicates the task pointer is invalid.

NU_INVALID_DELETE—indicates the task is not in a *Finished* or *Terminated* state.

Change task priority

This API assigns a new priority to a task. It is not needed with Nucleus SE as tasks' priorities are fixed.

Service call prototype:

```
OPTION NU_Change_Priority(
NU_TASK *task,
OPTION new_priority);
```

Parameters:

> `task`—pointer to task control block.
>
> `new_priority`—specifies a priority from 0 to 255.

Returns:

> The task's previous priority value

Change task preemption

This API call changes the preemption posture of the currently executing task. It is not needed with Nucleus SE, as a simpler scheduling scheme is utilized.

Service call prototype:

```
OPTION NU_Change_Preemption(OPTION preempt);
```

Parameters:

> `preempt`—new   preemption   posture: `NU_PREEMPT`   or
> `NU_NO_PREEMPT`

Returns:

> The task's previous preemption posture

Change task time slice

This API call changes the time slice of the specified task. It is not needed with Nucleus SE, as task time slice durations are fixed.

Service call prototype:

```
UNSIGNED NU_Change_Time_Slice(
NU_TASK *task,
UNSIGNED time_slice);
```

Parameters:

> `task`—pointer to the task control block;
>
> `time_slice`—the maximum amount of timer ticks that can expire while executing this task; a value of zero in this field disables time slicing for this task.

Returns:

> The task's previous time slice value

Terminate task

This API call terminates the specified task. It is not needed with Nucleus SE, as the *Terminated* state is not supported.

Service call prototype:

```
STATUS NU_Terminate_Task(NU_TASK *task);
```

Parameters:

task—pointer to the task control block

Returns:

NU_SUCCESS—indicates successful completion of the service.

NU_INVALID_TASK—indicates the task pointer is invalid.

Compatibility with Nucleus RTOS

With all aspects of Nucleus SE, it was my goal to maintain as high a level of application code compatibility with Nucleus RTOS as possible. Tasks are no exception and, from a user's perspective, they are implemented in much the same way as in Nucleus RTOS. There are areas of incompatibility, which have come about where I determined that such an incompatibility would be acceptable, given that the resulting code is easier to understand, or, more likely, could be made more memory efficient. Otherwise, Nucleus RTOS API calls may be almost directly mapped onto Nucleus SE calls. Chapter 20, *Using Nucleus SE*, includes further information on using Nucleus SE for users of Nucleus RTOS.

Object identifiers

In Nucleus RTOS, all objects are described by a data structure—a control block—which has a specific data type. A pointer to this control block serves as an identifier for the task. In Nucleus SE, I decided that a different approach was needed for memory efficiency, and all kernel objects are described by a number of tables in RAM and/or ROM. The size of these tables is determined by the number of each object type that is configured. The identifier for a specific object is simply an index into those tables. So, I have defined **NUSE_TASK** as being equivalent to **U8**; a variable—not a pointer—of this type then serves as the task identifier. This is a small incompatibility, which is easily handled if code is ported to or from Nucleus RTOS. Object identifiers are normally just stored and passed around and not operated on in any way.

Nucleus RTOS also supports naming of tasks. These names are only used for target-based debug facilities. I omitted them from Nucleus SE to save memory.

Task states

In Nucleus RTOS, tasks may in one of a number of states: *Executing, Ready, Suspended* (which may be indefinite, sleep, or API call blocked), *Terminated*, or *Finished*.

Nucleus SE always supports *Executing* and *Ready*. All three variants of *Suspended* are optionally supported. *Terminated* and *Finished* are not supported. There are no API calls appertaining to task termination. A task's outer function must never return, either explicitly or by "falling off the end" (this would result in the *Finished* state in Nucleus RTOS).

Unimplemented API calls

Nucleus RTOS supports 16 service calls to work with tasks. Of these, seven are not implemented in Nucleus SE. Details of these and of the decision to omit them were outlined above.

7

Partition memory

Memory partitions were introduced in Chapter 3, *RTOS Services and Facilities*, where a comparison with the standard C `malloc()` function was made. A partition is a memory area, obtained from a partition pool. This offers a flexible means for tasks to obtain and release data storage in a deterministic and reliable fashion.

Using partitions

In Nucleus SE, partition pools are configured at build time. There may be a maximum of 16 partition pools configured for an application. If no partition pools are configured, no data structures or service call code appertaining to partition pools are included in the application.

A partition pool is simply an area of memory, which is divided into a specific number of fixed-sized partitions. The user has complete control over the size and number of partitions in each pool. Tasks may request to be allocated memory partitions and receive a pointer to the storage area. It is the task's responsibility not to write data outside of the partition. The partition may be deallocated by any task by passing the pointer to an application program interface (API) function. Requesting the allocation of a partition, when no more are available, may result in an error or task suspension, depending on options selected in the API call and the Nucleus SE configuration.

Configuring memory partitions

Number of partition pools

As with most aspects of Nucleus SE, the configuration of partition pools is primarily controlled by `#define` statements in `nuse_config.h`. The key setting is `NUSE_PARTITION_POOL_NUMBER`, which determines how many partition pools are configured for the application. The default setting is 0 (i.e., no partition pools are in use) and you can set it to any value up to 16. An

Embedded RTOS Design. DOI: https://doi.org/10.1016/B978-0-12-822851-7.00007-2

erroneous value will result in a compile time error, which is generated by a test in `nuse_config_check.h` (this is included into `nuse_config.c` and hence compiled with this module), resulting in a `#error` statement being compiled.

Choosing a nonzero value is the "master enable" for partition pools. This results in some data structures being defined and sized accordingly. Data structures in ROM require initializing to suitable values that characterize each partition pool. There is more information on data structures later in this chapter. It also activates the API enabling settings.

API enables

Every API function (service call) in Nucleus SE has an enabling `#define` symbol in `nuse_config.h`. For partition pools, these are:

```
NUSE_PARTITION_ALLOCATE
NUSE_PARTITION_DEALLOCATE
NUSE_PARTITION_POOL_INFORMATION
NUSE_PARTITION_POOL_COUNT
```

By default, all of these are set to FALSE, thus disabling each service call and inhibiting the inclusion of any implementation code. To configure partition pools for an application, you need to select the API calls that you want to use and set their enabling symbols to TRUE.

Here is an extract from the default `nuse_config.h` file:

```
/*** Partition pools ***/

#define NUSE_PARTITION_POOL_NUMBER      0      /* Number of
                                                  partition pools
                           in the system - 0-16 */

#define NUSE_PARTITION_ALLOCATE         FALSE    /* Service call
                                                   enabler */
#define NUSE_PARTITION_DEALLOCATE       FALSE    /* Service call
                                                   enabler */
#define NUSE_PARTITION_POOL_INFORMATION FALSE    /* Service
                                                   call enabler */
#define NUSE_PARTITION_POOL_COUNT       FALSE    /* Service call
                                                   enabler */
```

A compile time error will result if a partition pool API function is enabled and no partition pools are configured (except for `NUSE_Partition_Pool_Count()`, which is always permitted). If your code uses an API call, which has not been enabled, a link time error will result, as no implementation code will have been included in the application.

Partition pool service calls

Nucleus RTOS supports seven service calls, which appertain to partition pools that provide the following functionality:

Allocate a partition: `NU_Allocate_Partition()`. Implemented by `NUSE_Partition_Allocate()` in Nucleus SE.

Deallocate a partition: `NU_Deallocate_Partition()`. Implemented by `NUSE_Partition_Deallocate()` in Nucleus SE.

Provide information about a specified partition pool: `NU_Partition_Pool_Information()`. Implemented by `NUSE_Partition_Pool_Information()` in Nucleus SE.

Return a count of how many partition pools are (currently) configured for the application: `NU_Established_Partition_Pools()`. Implemented by `NUSE_Partition_Pool_Count()` in Nucleus SE.

Add a new partition pool to the application (create): `NU_Create_Partition_Pool()`. Not implemented in Nucleus SE.

Remove a partition pool from the application (delete): `NU_Delete_Partition_Pool()`. Not implemented in Nucleus SE.

Return pointers to all the partition pools (currently) in the application: `NU_Partition_Pool_Pointers()`. Not implemented in Nucleus SE.

The implementation of each of these service calls is examined in detail.

It may be noted that no reset function is provided (in either Nucleus RTOS or Nucleus SE). This is intentional. It is very common practice for one task to allocate a partition and pass a pointer to another task (which will probably deallocate it later). If a partition pool were to be reset, all the partitions would be marked as unused, but there is no means of tracking and notifying all tasks that may be using the partitions.

Partition allocation and deallocation services

The fundamental operations, which can be performed on a partition pool, are allocating a partition (i.e., marking a partition as used and returning its address) from it and deallocating a partition (i.e., marking a partition as unused). Nucleus RTOS and Nucleus SE each provide two basic API calls for these operations, which will be discussed here.

Allocating a partition

The Nucleus RTOS API call for allocating a partition is very flexible, enabling you to suspend indefinitely, or with a timeout, if the operation cannot be completed immediately; that is, you try to allocate a partition from a pool in which all partitions are currently allocated. Nucleus SE provides the same service, except task suspend is optional and timeout is not implemented.

Nucleus RTOS API call for partition allocation

Service call prototype:

```
STATUS NU_Allocate_Partition(
NU_PARTITION_POOL *pool,
VOID **return_pointer,
UNSIGNED (suspend);
```

Parameters:

pool—pointer to the partition pool to be utilized;

return_pointer—a pointer to a pointer that will receive the address of the allocated partition;

suspend—specification for task suspend; may be NU_NO_SUSPEND or NU_SUSPEND or a timeout value in ticks (1–4,294,967,293).

Returns:

NU_SUCCESS—The call was completed successfully.

NU_NO_PARTITION—No partitions are available.

NU_INVALID_POOL—The partition pool pointer is invalid.

NU_INVALID_POINTER—The data return pointer is NULL.

NU_INVALID_SUSPEND—Suspend was attempted from a non-task thread.

NU_TIMEOUT—No partition is available even after suspending for the specified timeout period.

NU_POOL_DELETED—Partition pool was deleted, while the task was suspended.

Nucleus SE API call for allocating

This API call supports the key functionality of the Nucleus RTOS API.

Service call prototype:

```
STATUS NUSE_Partition_Allocate(
NUSE_PARTITION_POOL pool,
ADDR *return_pointer,
U8 suspend);
```

Parameters:

pool—the index (ID) of the partition pool to be utilized;

return_pointer—a pointer to a variable of type **ADDR**, which will receive the address of the allocated partition;

suspend—specification for task suspend; may be **NUSE_NO_SUSPEND** or **NUSE_SUSPEND**.

Returns:

NUSE_SUCCESS—The call was completed successfully.

NUSE_NO_PARTITION—No partitions are available.

NUSE_INVALID_POOL—The partition pool index is invalid.

NUSE_INVALID_POINTER—The data return pointer is **NULL**.

NUSE_INVALID_SUSPEND—Suspend was attempted from a non-task thread or when blocking API calls were not enabled.

Nucleus SE implementation of partition allocation

The bulk of the code of the **NUSE_Partition_Allocate()** API function—after parameter checking—is selected by conditional compilation, dependent on whether support for blocking (task suspend) API calls is enabled. We will look at the two variants separately here.

If blocking is not enabled, the code for this API call is quite straightforward:

```
if (NUSE_Partition_Pool_Partition_Number[pool] ==
NUSE_Partition_Pool_Partition_Used[pool])
{
```

```
      return_value = NUSE_NO_PARTITION;      /* no free partitions */
   }
   else
   {
     /* point to status byte of 1st partition */
     ptr = (U8 *)NUSE_Partition_Pool_Data_Address[pool];

     while (*ptr != 0)      /* find an unused partition */
     {
       ptr += NUSE_Partition_Pool_Partition_Size[pool] + 1;
     }
     *ptr = 0x80 | pool; /* mark used */
                   /* need pool number for deallocation */
     NUSE_Partition_Pool_Partition_Used[pool] ++;
     *return_pointer = ptr + 1;
     return_value = NUSE_SUCCESS;
   }
```

First, the availability of free partitions is checked and an error returned (**NUSE_ NO_PARTITION**) if there is none. The partitions are then stepped through, inspecting the first byte for a zero value (meaning unused). When this is found, an in-use flag is written, incorporating the index of the partition pool (see "Deallocating a sartition" section), and a pointer to the next byte (the start of the real data area) returned. For an explanation of the partition pool data structures, see "Data structures" section.

If blocking is enabled, the code for this API call is a little more complex:

```
do
{
  if (NUSE_Partition_Pool_Partition_Number[pool] ==
    NUSE_Partition_Pool_Partition_Used[pool])
  {
    if (suspend == NUSE_NO_SUSPEND)
    {
      return_value = NUSE_NO_PARTITION;  /* no free partitions */
    }
    else
    {                      /* block task */
      NUSE_Partition_Pool_Blocking_Count[pool] ++;
      NUSE_Suspend_Task(NUSE_Task_Active, (pool << 4) |
        NUSE_PARTITION_SUSPEND);
      return_value =
      NUSE_Task_Blocking_Return[NUSE_Task_Active];
      if (return_value != NUSE_SUCCESS)
      {
        suspend = NUSE_NO_SUSPEND;
```

```
            }
         }
      }
      else
      {
         /* point to status byte of 1st partition */
         ptr = (U8 *)NUSE_Partition_Pool_Data_Address[pool];
         while ( *ptr!=0)      /* find an unused partition */
         {
             ptr +=NUSE_Partition_Pool_Partition_Size[pool] + 1;
         }
         *ptr = 0x80 | pool; /* mark used */
                       /* need pool number for deallocation */
         NUSE_Partition_Pool_Partition_Used[pool]++;
         *return_pointerptr + 1;
         return_value =NUSE_SUCCESS;
         suspend =NUSE_NO_SUSPEND;
      }
   } while (suspend ==NUSE_SUSPEND);
```

The code is enclosed in a do...while loop, which continues, while the parameter suspend has the value NUSE_SUSPEND.

If a partition is available, it is marked as being in use, as described earlier, and the data pointer is set up. The suspend variable is set to NUSE_NO_SUSPEND and the API call exits with NUSE_SUCCESS.

If no partitions are available and suspend is set to NUSE_NO_SUSPEND, the API call exits with NUSE_ NO_PARTITION. If suspend was set to NUSE_SUSPEND, the task is suspended. On return (i.e., when the task is woken up), if the return value is NUSE_SUCCESS, indicating that the task was woken because a partition has been deallocated, the code loops back to the top. Of course, since there is no reset API function for partition pools, this is the only possible reason for the task to be woken, but the process of checking NUSE_Task_Blocking_Return[] is retained for consistency with the handling of blocking with other object types.

Deallocating a partition

Deallocating a partition in Nucleus RTOS and Nucleus SE simply makes it available for allocation again. No check is made (or could be made) on whether any task is still using the partition; this is the responsibility of the applications programmer. Only the pointer to the data area is required in order to deallocate a partition.

Nucleus RTOS API call for partition deallocation

Service call prototype:

```
STATUS NU_Deallocate_Partition(VOID *partition);
```

Parameters:

partition—pointer to the data area (as returned by **NU_Allocate_Partition()**) of the partition to be deallocated.

Returns:

NU_SUCCESS—The call was completed successfully.

NU_INVALID_POINTER—The partition pointer is **NULL** or does not appear to point to a valid, in-use partition.

Nucleus SE API call for partition deallocation

This API call supports the key functionality of the Nucleus RTOS API.

Service call prototype:

```
STATUS NUSE_Partition_Deallocate(ADDR partition);
```

Parameters:

partition—pointer to the data area (as returned by **NUSE_Partition_Allocate()**) of the partition to be deallocated.

Returns:

NUSE_SUCCESS—The call was completed successfully.

NUSE_INVALID_POINTER—The partition pointer is **NULL** or does not appear to point to a valid, in-use partition.

Nucleus SE implementation of partition deallocation

Instead of being implemented in blocking and nonblocking forms, the **NUSE_Partition_Deallocate()** API function simply contains a conditionally compiled section that deals with task unblocking. This code implements the actual partition deallocation:

```
pool =*((U8*)partition-1) & 0x0f;     /* extract pool index
                                                number */
*((U8*)partition-1) = 0;            /* mark unused */
NUSE_Partition_Pool_Partition_Used[pool]--;   /* decrement
                                                     count */
return_value =NUSE_SUCCESS;
```

First the index of the partition pool is extracted from the partition status byte. Then, the status is set to "unused," the usage count decremented and success reported on return.

If task blocking is enabled, this code is included to take care of waking up any tasks waiting on the partition pool:

```
if (NUSE_Partition_Pool_Blocking_Count[pool] != 0)
{
  U8 index;          /* check whether a task is blocked */
                /* on this partition pool */

  NUSE_Partition_Pool_Blocking_Count[pool]--;
  for (index =0; index <NUSE_TASK_NUMBER; index++)
  {
    if ((LONIB(NUSE_Task_Status[index]) ==
      NUSE_PARTITION_SUSPEND)
      && (HINIB(NUSE_Task_Status[index]) ==pool))
    {
      NUSE_Task_Blocking_Return[index] =NUSE_SUCCESS;
      NUSE_Wake_Task(index);
      break;
    }
  }
}
```

If any tasks are blocked on allocation of a partition in this pool, the first in the table is woken.

Partition pool utility services

Nucleus RTOS has three API calls that provide utility functions associated with partition pools: return information about a partition pool, return number of partition pools in the application, and return pointers to all partition pools in the application. The first two of these are implemented in Nucleus SE.

Obtaining partition pool information

This service call obtains a selection of information about a partition pool. The Nucleus SE implementation differs from Nucleus RTOS in that it returns less information, as object naming and suspend ordering are not supported and task sus-pend may not be enabled.

Nucleus RTOS API call for partition pool information

Service call prototype:

```
STATUS NU_Partition_Pool_Information(
NU_PARTITION_POOL *pool,
CHAR *name,
VOID **start_address,
UNSIGNED *pool_size,
UNSIGNED *partition_size,
UNSIGNED *available,
UNSIGNED *allocated,
OPTION *suspend_type,
UNSIGNED *tasks_waiting,
NU_TASK **first_task);
```

Parameters:

pool—pointer to the partition pool about which informa-tion is being requested;

name—pointer to an 8-character destination area for the parti-tion pool's name; this includes space for the NULL terminator;

start_address—a pointer to a variable, which will receive a pointer to the start of the partition pool data area;

pool_size—a pointer to a variable, which will receive the size of the partition pool (in bytes);

partition_size—a pointer to a variable, which will receive the size of partitions in this pool;

available—a pointer to a variable, which will receive the number of partitions currently available in this pool;

allocated—a pointer to a variable, which will receive the number of partitions currently in use in this pool;

suspend_type—pointer to a variable for holding the task suspend type; valid task suspend types are NU_FIFO and NU_PRIORITY;

tasks_waiting—a pointer to a variable, which will receive the number of tasks suspended on this partition pool;

first_task—a pointer to a task pointer into which is placed the pointer of the first suspended task.

Returns:

> **NU_SUCCESS**—The call was completed successfully.
>
> **NU_INVALID_POOL**—The partition pool pointer is not valid.

Nucleus SE API call for partition pool information

This API call supports the key functionality of the Nucleus RTOS API.

Service call prototype:

```
STATUS NUSE_Partition_Pool_Information(
NUSE_PARTITION_POOL pool,
ADDR *start_address,
U32 *pool_size,
U16 *partition_size,
U8 *available,
U8 *allocated,
U8 *tasks_waiting,
NUSE_TASK *first_task);
```

Parameters:

> **pool**—the index of the partition pool about which information is being requested;
>
> **start_address**—a pointer to a variable, which will receive a pointer to the start of the partition pool data area;
>
> **pool_size**—a pointer to a variable, which will receive the size of the partition pool (in bytes);
>
> **partition_size**—a pointer to a variable, which will receive the size of partitions in this pool;
>
> **available**—a pointer to a variable, which will receive the number of partitions currently available in this pool;
>
> **allocated**—a pointer to a variable, which will receive the number of partitions currently in use in this pool;
>
> **tasks_waiting**—a pointer to a variable which will receive the number of tasks suspended on this partition pool (nothing returned if task suspend is disabled);
>
> **first_task**—a pointer to a variable of type **NUSE_TASK**, which will receive the index of the first suspended task (nothing returned if task suspend is disabled).

Returns:

NUSE_SUCCESS—The call was completed successfully.

NUSE_INVALID_POOL—The partition pool index is not valid.

NUSE_INVALID_POINTER—One or more of the pointer parameters is invalid.

Nucleus SE implementation of partition pool information

The implementation of this API call is quite straightforward:

```
STATUS NUSE_Partition_Pool_Information(
  NUSE_PARTITION_POOL pool,
  ADDR *start_address, U32 *pool_size, U16 *partition_size,
  U8 *available, U8 *allocated, U8 *tasks_waiting,
  NUSE_TASK *first_task)
{
  #if NUSE_API_PARAMETER_CHECKING
    if (pool >= NUSE_PARTITION_POOL_NUMBER)
    {
      return NUSE_INVALID_POOL;
    }

    if ((start_address == NULL) || (pool_size == NULL) ||
        (partition_size == NULL) || (available == NULL) ||
        (allocated == NULL) || (tasks_waiting == NULL) ||
        (first_task == NULL))
    {
      return NUSE_INVALID_POINTER;
    }
  #endif

  NUSE_CS_Enter();

  *start_address = NUSE_Partition_Pool_Data_Address[pool];
  *pool_size = NUSE_Partition_Pool_Partition_Number[pool] *
      NUSE_Partition_Pool_Partition_Size[pool];
  *partition_size = NUSE_Partition_Pool_Partition_Size[pool];
  *available = NUSE_Partition_Pool_Partition_Number[pool] -
      NUSE_Partition_Pool_Partition_Used[pool];
  *allocated = NUSE_Partition_Pool_Partition_Used[pool];

  #if NUSE_BLOCKING_ENABLE

    *tasks_waiting = NUSE_Partition_Pool_Blocking_Count[pool];
    if (NUSE_Partition_Pool_Blocking_Count[pool] != 0)
    {
      U8 index;

      for (index = 0; index < NUSE_TASK_NUMBER; index++)
```

```
      {
        if ((LONIB(NUSE_Task_Status[index]) ==
          NUSE_PARTITION_SUSPEND)
          && (HINIB(NUSE_Task_Status[index]) == pool))
        {
          *first_task = index;
          break;
        }
      }
    }
    else
    {
      *first_task = 0;
    }

  #else

    *tasks_waiting = 0;
    *first_task = 0;

  #endif

  NUSE_CS_Exit();

  return NUSE_SUCCESS;
}
```

The function returns the partition pool status. Then, if block-ing API calls is enabled, the number of waiting tasks and the index of the first one are returned (otherwise these two para-meters are set to 0).

Obtaining the number of partition pools

This service call returns the number of partition pools con-figured in the application. Whilst in Nucleus RTOS this will vary over time and the returned value will represent the current number of pools, in Nucleus SE the value returned is set at build time and cannot change.

Nucleus RTOS API call for number of partition pools
Service call prototype:

```
UNSIGNED NU_Established_Partition_Pools(VOID);
```

Parameters:

    None

Returns:

The number of created partition pools in the application

Nucleus SE API call for number of partition pools

This API call supports the key functionality of the Nucleus RTOS API.

Service call prototype:

```
U8 NUSE_Partition_Pool_Count(void);
```

Parameters:

None

Returns:

The number of configured partition pools in the application

Nucleus SE implementation of number of partition pools

The implementation of this API call is almost trivially simple: the value of the `#define` symbol `NUSE_PARTITION_POOL_NUMBER` is returned.

Data structures

Partition pools utilize a number of data structures—in ROM and RAM—which, like other Nucleus SE objects, are a series of tables, included and dimensioned according to the number of pools configured and options selected.

I strongly recommend that application code does not access these data structures directly, but uses the provided API functions. This avoids incompatibility with future versions of Nucleus SE and unwanted side-effects and simplifies porting of an application to Nucleus RTOS. The details of data structures are included here to facilitate easier understanding of the working of the service call code and for debugging.

Kernel RAM data

These data structures are:

`NUSE_Partition_Pool_Partition_Used[]`—This is an array of type `U8`, with one entry for each configured partition pool, which contains a count of the number of partitions currently in use.

`NUSE_Partition_Pool_Blocking_Count[]`—This type `U8` array contains the counts of how many tasks are blocked on each partition pool. This array only exists if blocking API call support is enabled.

These data structures are all initialized to zeros by `NUSE_Init_Partition_Pool()` when Nucleus SE starts up. This is logical, as it renders every partition in every pool unused. A full description of Nucleus SE start-up procedures may be found in Chapter 17, *Nucleus SE Initialization and Start-Up*.

Here are the definitions of these data structures in `nuse_init.c` file.

```
RAM U8 NUSE_Partition_Pool_Partition_Used
[NUSE_PARTITION_POOL_NUMBER];

#if NUSE_BLOCKING_ENABLE

RAM U8 NUSE_Partition_Pool_Blocking_Count
[NUSE_PARTITION_POOL_NUMBER];

#endif
```

User RAM

It is the user's responsibility to provide an area of RAM for data storage for each configured partition pool. The size of this RAM area must accommodate the size and number of partitions configured (see "ROM data" section), with an additional byte for each partition in the pool. Each partition has a data area preceded by a single status byte.

ROM data

These data structures are:

`NUSE_Partition_Pool_Data_Address[]`—This is an array of type `ADDR`, with one entry for each configured partition pool, which contains the address of the start of the data storage area.

`NUSE_Partition_Pool_Partition_Number[]`—This is an array of type `U8`, with one entry for each configured partition pool, which contains the number of partitions in the pool.

`NUSE_Partition_Pool_Partition_Size[]`—This is an array of type `U16`, with one entry for each configured partition pool, which contains the partition size for the pool.

These data structures are all declared and initialized (statically, of course) in `nuse_config.c`, this:

```
ROM ADDR NUSE_Partition_Pool_Data_Address
[NUSE_PARTITION_POOL_NUMBER] =
{
  /* address of partition pools ------ */
};

ROM U8 NUSE_Partition_Pool_Partition_Number
[NUSE_PARTITION_POOL_NUMBER] =
{
  /* number of partitions in each pool ------ */
};

ROM U16 NUSE_Partition_Pool_Partition_Size
[NUSE_PARTITION_POOL_NUMBER] =
{
  /* partition sizes ------ */
  /* in bytes */
};
```

Partition pool data footprint

Like all kernel objects in Nucleus SE, the amount of data memory required for partition pools is readily predictable.

The ROM data footprint (in bytes) for all the partition pools in an application may be computed thus:

```
NUSE_PARTITION_POOL_NUMBER * (sizeof(ADDR) + 2)
```

The kernel RAM data footprint for all the partition pools in an application, when blocking API calls is enabled, is simply two bytes per partition pool. If blocking is disabled, only one byte is required.

The user RAM size for each partition pool varies, but, as mentioned earlier, may be computed for a pool of index `n` thus:

```
NUSE_Partition_Pool_Partition_Number[n] *
(NUSE_Partition_Pool_Partition_Size[n] + 1)
```

Unimplemented API calls

Three partition pool API calls found in Nucleus RTOS are not implemented in Nucleus SE:

Create partition pool

This API call creates a partition pool. It is not needed with Nucleus SE, as tasks are created statically.

Service call prototype:

```
STATUS NU_Create_Partition_Pool(
NU_PARTITION_POOL *pool,
CHAR *name,
VOID *start_address,
UNSIGNED pool_size,
UNSIGNED partition_size,
OPTION suspend_type);
```

Parameters:

pool—pointer to a user-supplied partition pool control block; this will be used as a "handle" for the partition pool in other API calls.

name—pointer to a 7-character, NULL-terminated name for the partition pool;

start_address—specifies the starting address for the partition pool memory area.

pool_size—the total number of bytes in the memory area.

partition_size—the number of bytes for each partition in the pool. There is a small amount of memory "overhead" associated with each partition; this overhead is required by the two data pointers used.

suspend_type—specifies how tasks suspend on the partition pool; valid options for this parameter are NU_FIFO and NU_PRIORITY.

Returns:

NU_SUCCESS—indicates successful completion of the service.

NU_INVALID_POOL—indicates the partition pool control block pointer is NULL.

NU_INVALID_MEMORY—indicates the memory area specified by start_ address is NULL.

NU_INVALID_SIZE—indicates the partition size is either 0 or larger than the total partition memory area.

NU_INVALID_SUSPEND—indicates the suspend_type parameter is invalid.

Delete partition pool

This API call deletes a previously created partition pool. It is not needed with Nucleus SE, as tasks are created statically and cannot be deleted.

Service call prototype:

```
STATUS NU_Delete_Partition_Pool(NU_PARTITION_POOL *pool);
```

Parameters:

pool—pointer to partition pool control block.

Returns:

NU_SUCCESS—indicates successful completion of the service.

NU_INVALID_POOL—indicates the partition pool pointer is invalid.

Partition pool pointers

This API call builds a sequential list of pointers to all partition pools in the system. It is not needed with Nucleus SE, as tasks are identified by a simple index, not a pointer, and it would be redundant.

Service call prototype:

```
UNSIGNED NU_Partition_Pool_Pointers(
NU_PARTITION_POOL **pointer_list,
UNSIGNED maximum_pointers);
```

Parameters:

pointer_list—pointer to an array of **NU_PARTITION_POOL** pointers; this array will be filled with pointers to established partition pools in the system.

maximum_pointers—the maximum number of pointers to place in the array.

Returns:

The number of **NU_PARTITION_POOL** pointers placed into the array.

Compatibility with Nucleus RTOS

With all aspects of Nucleus SE, it was my goal to maintain as high a level of applications code compatibility with Nucleus RTOS as possible. Partition pools are no exception and, from a user's perspective, they are implemented in much the same way as in Nucleus RTOS. There are areas of incompatibility, which have come about where I determined that such an incompatibility would be acceptable, given that the resulting code is easier to understand, or, more likely, could be made more memory efficient. Otherwise, Nucleus RTOS API calls may be almost directly mapped onto Nucleus SE calls. Chapter 20, *Using Nucleus SE*, provides further information on using Nucleus SE for users of Nucleus RTOS.

Object identifiers

In Nucleus RTOS, all objects are described by a data structure—a control block—which has a specific data type. A pointer to this control block serves as an identifier for the partition pool. In Nucleus SE, I decided that a different approach was needed for memory efficiency, and all kernel objects are described by a number of tables in RAM and/or ROM. The size of these tables is determined by the number of each object type that is configured. The identifier for a specific object is simply an index into those tables. So, I have defined **NUSE_PARTITION_POOL** as being equivalent to **U8**; a variable—not a pointer—of this type then serves as the task identifier. This is a small incompatibility, which is easily handled if code is ported to or from Nucleus RTOS. Object identifiers are normally just stored and passed around and not operated on in any way.

Nucleus RTOS also supports naming of partition pools. These names are only used for target-based debug facilities. I omitted them from Nucleus SE to save memory.

Partition size and number

In Nucleus RTOS a partition pool is configured in terms of the overall pool size and the partition size (which carries a two-pointer overhead); these parameters are expressed as type **UNSIGNED**, which is probably 32 bits. In Nucleus SE a partition pool is configured in terms of partition size (to which an additional byte of overhead is added) and number of partitions; these parameters are expressed as a **U16** and **U8**, respectively.

Unimplemented API calls

Nucleus RTOS supports seven service calls to work with partition pools. Three of these service calls are not implemented in Nucleus SE. Details of these and of the decision to omit them were outlined earlier.

8

Signals

In this chapter, I am going to look at signals, which are the simplest method of intertask communication supported by Nucleus SE. They provide a very low cost by means of passing simple messages between tasks.

Using signals

Signals are different from all the other types of kernel objects, in that they are not autonomous—signals are associated with tasks and have no independent existence. If signals are configured for an application, each task has a set of eight signal flags.

Any task can set the signals of another task. Only the owner task can read the signals. The read is destructive—that is, the signals are cleared by the process of reading. No other task can read or clear a task's signals.

There is a facility in Nucleus RTOS that enables a task to nominate a function that is run when another task sets one or more of its signal flags. This is somewhat analogous to an interrupt service routine. This capability is not supported in Nucleus SE; tasks need to interrogate their signal flags explicitly.

Configuring signals

As with most aspects of Nucleus SE, the configuration of signals is primarily controlled by `#define` statements in `nuse_config.h`. The key setting is `NUSE_SIGNAL_SUPPORT`, which enables the facility (for all tasks in the application). There is no question of specifying the number of signals—there is simply a set of eight flags for each task in the application.

Setting this enable parameter is the "master enable" for signals. This causes a data structure to be defined and sized accordingly, of which more later in this chapter. It also activates the application program interface (API) enabling settings.

Embedded RTOS Design. DOI: https://doi.org/10.1016/B978-0-12-822851-7.00008-4

Application program interface enables

Every API function (service call) in Nucleus SE has an enabling #define symbol in `nuse_config.h`. For signals, these are:

```
NUSE_SIGNALS_SEND
NUSE_SIGNALS_RECEIVE
```

By default, both of these are set to FALSE, thus disabling each service call and inhibiting the inclusion of any implementation code. To configure signals for an application, you need to select the API calls that you want to use and set their enabling symbols to TRUE.

Here is an extract from the default `nuse_config.h` file:

```
#define NUSE_SIGNAL_SUPPORT FALSE /* Enables support for signals */

#define NUSE_SIGNALS_SEND FALSE /* Service call enabler */
#define NUSE_SIGNALS_RECEIVE FALSE /* Service call enabler */
```

It will result in a compile-time error if a signal API function is enabled and the signal facility has not been enabled. If your code uses an API call, which has not been enabled, it will result in a link time error, as no implementation code will have been included in the application. Of course, the enabling of the two API functions is somewhat redundant, as there would be no point in enabling signal support and not having these APIs available. The enables are included for compatibility with other Nucleus SE features.

Signals service calls

Nucleus RTOS supports four service calls that appertain to signals, which provide the following functionality:

Send signals to a specified task. Implemented by `NUSE_Signals_Send()` in Nucleus SE.

Receive signals. Implemented by `NUSE_Signals_Receive()` in Nucleus SE.

Register a signal handler. Not implemented in Nucleus SE.

Enable/disable (control) signals. Not implemented in Nucleus SE.

The implementation of each of these service calls will now be examined in detail.

Signals send and receive services

The fundamental operations, which can be performed on a task's set of signals, are sending data to it (which may be done by any task) and reading data from it (and thus clearing the data, which may only be done by the owner). Nucleus RTOS and Nucleus SE each provides two basic API calls for these operations, that will be discussed here.

Since signal flags are bits, they are best visualized as binary numbers. As standard C does not historically support a representation of binary constants (only octal and hexadecimal), the Nucleus SE distribution includes a useful header file—`nuse_binary.h`—which contains `#define` symbols of the form `b01010101` for all 256 8-bit values. Here is an extract from the `nuse_binary.h` file:

```
#define b00000000 ((U8) 0x00)
#define b00000001 ((U8) 0x01)
#define b00000010 ((U8) 0x02)
#define b00000011 ((U8) 0x03)
#define b00000100 ((U8) 0x04)
#define b00000101 ((U8) 0x05)
```

Sending signals

Any task may send signals to any other task in the application. Sending signals involves setting one or more of the signal flags. This is an OR operation that has no effect on flags set previously.

Nucleus RTOS API call for sending signals

Service call prototype:

```
STATUS NU_Send_Signals(
NU_TASK *task,
UNSIGNED signals);
```

Parameters:

  `task`—pointer to control block of the task that owns the signal flags to be set

  `signals`—the value of the signal flags to be set

Returns:

  `NU_SUCCESS`—the call was completed successfully

  `NU_INVALID_TASK`—the task pointer is invalid

Nucleus SE API call for sending signals

This API call supports the key functionality of the Nucleus RTOS API.

Service call prototype:

```
STATUS NUSE_Signals_Send(
NUSE_TASK task,
U8 signals);
```

Parameters:

task—the index (ID) of the task that owns the signal flags to be set

signals—the value of the signal flags to be set

Returns:

NUSE_SUCCESS—the call was completed successfully

NUSE_INVALID_TASK—the task ID is invalid

Nucleus SE implementation of sending signals

Here is the complete code for the **NUSE_Signals_Send()** function:

```
STATUS NUSE_Signals_Send(NUSE_TASK task, U8 signals)
{
  #if NUSE_API_PARAMETER_CHECKING
    if (task >= NUSE_TASK_NUMBER)
      {
        return NUSE_INVALID_TASK;
      }
  #endif
    NUSE_CS_Enter();
    NUSE_Task_Signal_Flags[task] |= signals;
    NUSE_CS_Exit();

  return NUSE_SUCCESS;
}
```

The code is very simple. After any parameter checking, the signal values are ORed into the specified task's signal flags. Task blocking is not relevant to signals.

Receiving signals

A task may only read its own set of signal flags. The process of reading them is destructive; that is, it also results in the flags being cleared.

Nucleus RTOS API call for receiving signals

Service call prototype:

```
UNSIGNED NU_Receive_Signals(VOID);
```

Parameters:

None

Returns:

The signal flags value

Nucleus SE API call for receiving signals

This API call supports the key functionality of the Nucleus RTOS API.

Service call prototype:

```
U8 NUSE_Signals_Receive(void);
```

Parameters:

None

Returns:

The signal flag value

Nucleus SE implementation of receiving signals

Here is the complete code for the `NUSE_Signals_Receive()` function:

```
U8 NUSE_Signals_Receive(void)
{
    U8 signals;

    NUSE_CS_Enter();
    signals = NUSE_Task_Signal_Flags[NUSE_Task_Active];
    NUSE_Task_Signal_Flags[NUSE_Task_Active] = 0;
    NUSE_CS_Exit();

    return signals;
}
```

The code is very simple. The flag value is copied, the original value cleared and the copy returned by the API function. Task blocking is not relevant to signals.

Data structures

Since signals are not stand-alone objects, their memory usage really appertains to the task by which they are owned, but some notes are included for completeness. Signals utilize a single data structure—in random-access memory (RAM)—which, like other Nucleus SE objects, is a table, dimensioned according to the number of tasks configured and only included if signal support was selected.

I strongly recommend that application code does not access this data structure directly, but uses the provided API functions. This avoids incompatibility with future versions of Nucleus SE and unwanted side effects and simplifies porting of an application to Nucleus RTOS. The details of data structures are included here to facilitate easier understanding of the working of the service call code and for debugging.

RAM data

The data structure is:

NUSE_Task_Signal_Flags[]—This is an array of type u8, with one entry for each configured task and is where the signal flags are stored.

This data structure is initialized to zeros by NUSE_Init_Task() when Nucleus SE starts up.

ROM data

There are no read-only memory (ROM) data structures associated with signals.

Signal data footprint

Like all kernel objects in Nucleus SE, the amount of data memory required for signals is readily predictable.

The ROM data footprint for all the signals in an application is 0.

The RAM data footprint (in bytes) for all the signals in an application is simply the number of configured tasks (NUSE_TASK_NUMBER). Again, these data really "belong" to the tasks and is included in Chapter 6, *Tasks*.

Unimplemented API calls

Two signal API calls found in Nucleus RTOS are not implemented in Nucleus SE:

Register a signal handler

This API call establishes a signal handling routine (function) for the current task. It is not needed with Nucleus SE, as signal handlers are not supported.

Service call prototype:

```
STATUS NU_Register_Signal_Handler
(VOID(*signal_handler)(UNSIGNED));
```

Parameters:

`signal_handler`—function to be called whenever signals are received

Returns:

`NU_SUCCESS`—successful completion of the service

`NU_INVALID_POINTER`—the signal handler pointer is `NULL`

The signal handler is executed in the context of the current task.

Control (enable/disable) signals

This service enables and/or disables signals for the current task. There are 32 signals available for each task. Each signal is represented by a bit in `signal_enable_mask`. Setting a bit in `signal_enable_mask` enables the corresponding signal, while clearing a bit disables the corresponding signal.

Service call prototype:

```
UNSIGNED NU_Control_Signals(UNSIGNED enable_signal_mask);
```

Parameters:

`enable_signal_mask`—bit pattern representing valid signals

Returns:

The previous signal enables/disables mask.

Compatibility with Nucleus RTOS

With all aspects of Nucleus SE, it was my goal to maintain as high as a level of application code compatibility with Nucleus RTOS as possible. Signals are no exception and, from a user's perspective, they are implemented in much the same way as in Nucleus RTOS. There are areas of incompatibility, which have come about where I determined that such an incompatibility would be acceptable, given that the resulting code is easier to understand, or, more likely, could be made more memory efficient. Otherwise, Nucleus RTOS API calls may be almost directly mapped onto Nucleus SE calls.

Signal handlers

In Nucleus SE, signal handlers were not implemented to maintain simplicity.

Signal availability and number

In Nucleus RTOS, tasks each have 32 signal flags. In Nucleus SE, I decided to reduce this to eight signal flags per task, as this is likely to be sufficient for simpler applications and saves RAM. Signals may be disabled entirely, if they are not required.

Unimplemented API calls

Nucleus RTOS supports four service calls to work with signals. Of these, two are not implemented in Nucleus SE. Details of these and of the decision to omit them may be found in the section *Unimplemented API calls* earlier in this chapter.

9

Event flag groups

Event flag groups were introduced in Chapter 3, RTOS Services and Facilities. In Nucleus SE, they are somewhat similar to signals, but with greater flexibility. They provide a low cost, but flexible, means of passing simple messages between tasks.

Using event flags

In Nucleus SE, event flags are configured at build time. There may be a maximum of 16 event flag groups configured for an application. If no event flag groups are configured, no data structures or service call code appertaining to event flag groups are included in the application.

An event flag group is simply a set of eight 1-bit flags, access to which is controlled so that it may be safely utilized by multiple tasks. One task can set or clear any combination of event flags. Another task may read the event flag group at any time or may wait (polling or suspended) for a specific pattern of flags.

Configuring event flag groups

Number of event flag groups

As with most aspects of Nucleus SE, the configuration of event flag groups is primarily controlled by `#define` statements in `nuse_config.h`. The key setting is `NUSE_EVENT_GROUP_NUMBER`, which determines how many event flag groups are configured for the application. The default setting is 0 (i.e., no event flag groups are in use) and you can set it to any value up to 16. An erroneous value will result in a compile time error, which is generated by a test in `nuse_config_check.h` (this is included into `nuse_config.c` and hence compiled with this module), resulting in a `#error` statement being compiled.

Choosing a nonzero value is the "master enable" for event flag groups. This results in some data structures being defined and sized accordingly, of which more later in this chapter. It also activates the application program interface (API) enabling settings.

Embedded RTOS Design. DOI: https://doi.org/10.1016/B978-0-12-822851-7.00009-6

API enables

Every API function (service call) in Nucleus SE has an enabling #define symbol in nuse_config.h. For event flag groups, these are:

```
NUSE_EVENT_GROUP_SET
NUSE_EVENT_GROUP_RETRIEVE
NUSE_EVENT_GROUP_INFORMATION
NUSE_EVENT_GROUP_COUNT
```

By default, all of these are set to FALSE, thus disabling each service call and inhibiting the inclusion of any implementation code. To configure event flag groups for an application, you need to select the API calls that you want to use and set their enabling symbols to TRUE

Here is an extract from the default nuse_config.h file.

```
#define NUSE_EVENT_GROUP_NUMBER    0      /* Number of event groups
                                             in the system - 0-16 */

#define NUSE_EVENT_GROUP_SET     FALSE    /* Service call
                                             enabler */
#define NUSE_EVENT_GROUP_RETRIEVE    FALSE    /* Service call
                                                 enabler */
#define NUSE_EVENT_GROUP_INFORMATION    FALSE    /* Service
                                                    call enabler */
#define NUSE_EVENT_GROUP_COUNT    FALSE    /* Service call
                                             enabler */
```

A compile time error will result if an event flag group API function is enabled and no event flag groups are configured (except for NUSE_Event_Group_Count(), which is always permitted). If your code uses an API call, which has not been enabled, a link time error will result, as no implementation code will have been included in the application.

Event flag service calls

Nucleus RTOS supports seven service calls that appertain to event flags, which provide the following functionality:

Set event flags. Implemented by NUSE_Event_Group_Set() in Nucleus SE.

Retrieve event flags. Implemented by NUSE_Event_Group_Retrieve() in Nucleus SE.

Provide information about a specified event flag group. Implemented by `NUSE_Event_Group_Information()` in Nucleus SE.

Return a count of how many event flag groups are (currently) configured for the application. Implemented by `NUSE_Event_Group_Count()` in Nucleus SE.

Add a new event flag group to the application (create). Not implemented in Nucleus SE.

Remove an event flag group from the application (delete). Not implemented in Nucleus SE.

Return pointers to all the event flag groups (currently) in the application. Not implemented in Nucleus SE.

The implementation of each of these service calls will be examined in detail.

It may be noted that there is no reset function is provided (in either Nucleus RTOS or Nucleus SE). This is intentional. A reset implies a special condition is prevailing. For an event flag group, the only "special" condition would be all zeros, which can be set up with `NUSE_Event_Group_Set()`.

Event flag group set and retrieve services

The fundamental operations, which can be performed on an event flag group, are setting one or more flags and retrieving the current states of the flags. Nucleus RTOS and Nucleus SE each provide two basic API calls for these operations, which will be discussed here.

Since event flags are bits, they are best visualized as binary numbers. As standard C does not support a representation of binary constants (only octal and hexadecimal), the Nucleus SE distribution includes a useful header file—`nuse_binary.h`—which contains `#define` symbols of the form `b01010101` for all 256 8-bit values.

Setting event flags

The Nucleus RTOS API call for setting event flags is very flexible, enabling you to set and clear event flags using AND and OR operations. Nucleus SE provides the same service, except task suspend is optional.

Nucleus RTOS API call for flag setting

Service call prototype:

```
STATUS NU_Set_Events(
NU_EVENT_GROUP *group,
UNSIGNED event_flags,
OPTION operation);
```

Parameters:

group—pointer to the user-supplied event flag group control block;

event_flags—the bit value of the pattern of flags to be operated on;

operation—the operation to be performed; may be NU_OR (to set flags) or NU_AND (to clear flags).

Returns:

NU_SUCCESS—The call was completed successfully.

NU_INVALID_GROUP—The event flag group pointer is invalid.

NU_INVALID_OPERATION—The specified operation was not NU_OR or NU_AND.

Nucleus SE API call for flag setting

This API call supports the key functionality of the Nucleus RTOS API.

Service call prototype:

```
STATUS NUSE_Event_Group_Set(
NUSE_EVENT_GROUP group,
U8 event_flags,
OPTION operation);
```

Parameters:

group—the index (ID) of the event group in which flags are to be set/cleared;

event_flags—the bit value of the pattern of flags to be operated on;

operation—the operation to be performed; may be NUSE_OR (to set flags) or NUSE_AND (to clear flags).

Returns:

NUSE_SUCCESS—The call was completed successfully.

NUSE_INVALID_GROUP—The event flag group index is invalid.

NUSE_INVALID_OPERATION—The specified operation was not **NUSE_OR** or **NUSE_AND**.

Nucleus SE implementation of set event flags

The initial code of the **NUSE_Event_Group_Set()** API function—after parameter checking—is common, whether support for blocking (task suspend) API calls is enabled or not. The logic is quite simple:

```
NUSE_CS_Enter();

if (operation == NUSE_OR)
{
  NUSE_Event_Group_Data[group] |= event_flags;
}
else   /* NUSE_AND */
{
  NUSE_Event_Group_Data[group] &= event_flags;
}
```

The bit pattern, **event_flags**, is just ORed or ANDed into the specified event flag group.

The remaining code is only included if task blocking is enabled:

```
#if NUSE_BLOCKING_ENABLE

  while (NUSE_Event_Group_Blocking_Count[group] != 0)
  {
    U8 index;     /* check whether any tasks are blocked */
          /* on this event group */

    for (index = 0; index < NUSE_TASK_NUMBER; index++)
    {
      if ((LONIB(NUSE_Task_Status[index]) ==
          NUSE_EVENT_SUSPEND)
          && (HINIB(NUSE_Task_Status[index]) == group))
      {
        NUSE_Task_Blocking_Return[index] = NUSE_SUCCESS;
        NUSE_Task_Status[index] = NUSE_READY;
        break;
      }
    }
    NUSE_Event_Group_Blocking_Count[group]--;
  }
```

```
#if NUSE_SCHEDULER_TYPE == NUSE_PRIORITY_SCHEDULER
    NUSE_Reschedule(NUSE_NO_TASK);
#endif

#endif

NUSE_CS_Exit();

return NUSE_SUCCESS;
```

If any tasks are suspended on (retrieving from) this event flag group, they are all resumed. When they have the opportunity to execute, which depends on the scheduler, they can determine whether their conditions for return have been met—see "Retrieving event flags" section.

Retrieving event flags

The Nucleus RTOS API call for retrieving is very flexible, enabling you to suspend indefinitely, or with a timeout, if the operation cannot be completed immediately; that is, you try to retrieve a specific pattern of event flags, which does not represent the current state. Nucleus SE provides the same service, except task suspend is optional and timeout is not implemented.

Nucleus RTOS API call for flag retrieving

Service call prototype:

```
STATUS NU_Retrieve_Events(
NU_EVENT_GROUP *group,
UNSIGNED requested_events,
OPTION operation,
UNSIGNED *retrieved_events,
UNSIGNED suspend);
```

Parameters:

group—Pointer to the user-supplied event flag group control block.

requested_events—A bit pattern representing the flags to be retrieved.

operation—There are four operation options available: NU_AND, NU_AND_CONSUME, NU_OR, and NU_OR_CONSUME. NU_AND and NU_AND_CONSUME options indicate that all of the requested event flags are required. NU_OR and NU_OR_CONSUME options indicate that one or more of the requested event flags is sufficient. The CONSUME option automatically clears the event flags present on a successful request.

retrieved_events—A pointer to storage for the actual value of the retrieved event flags.

suspend—Specification for task suspend; may be NU_NO_SUSPEND or NUSE_SUSPEND or a timeout value in ticks (1–4,294,967,293).

Returns:

NU_SUCCESS—The call was completed successfully.

NU_NOT_PRESENT—Specified operation did not retrieve events (none present for NU_OR; not all present for NU_AND).

NU_INVALID_GROUP—The event flag group pointer is invalid.

NU_INVALID_OPERATION—The specified operation was invalid.

NU_INVALID_POINTER—The pointer to retrieved event flag storage is NULL.

NU_INVALID_SUSPEND—Suspend was attempted from a nontask thread.

NU_TIMEOUT—The requested event flag combination is not present even after the specified suspension timeout.

NU_GROUP_DELETED—The event flag group was deleted, while the task was suspended.

Nucleus SE API call for flag retrieving

This API call supports the key functionality of the Nucleus RTOS API.

Service call prototype:

```
STATUS NUSE_Event_Group_Retrieve(
NUSE_EVENT_GROUP group,
U8 requested_events,
OPTION operation,
U8 *retrieved_events,
U8 suspend);
```

Parameters:

group—the index (ID) of the event flag group to be utilized;

requested_events—a bit pattern representing the flags to be retrieved;

`operation`—specification for whether all or just some of the requested events need to be present; may be `NUSE_OR` (some flags) or `NUSE_AND` (all flags);

`retrieved_events`—a pointer to storage for the actual value of the retrieved event flags (same value as `requested_events` if operation was `NUSE_AND`);

`suspend`—specification for task suspend; may be `NUSE_NO_SUSPEND` or `NUSE_SUSPEND`.

Returns:

`NUSE_SUCCESS`—The call was completed successfully.

`NUSE_NOT_PRESENT`—Specified operation did not retrieve events (none present for `NUSE_OR`; not all present for `NUSE_AND`).

`NUSE_INVALID_GROUP`—The event flag group index is invalid.

`NUSE_INVALID_OPERATION`—The specified operation was not `NUSE_OR` or `NUSE_AND`.

`NUSE_INVALID_POINTER`—The pointer to retrieved event flag storage is `NULL`.

`NUSE_INVALID_SUSPEND`—Suspend was attempted from a nontask thread or when blocking API calls were not enabled.

Nucleus SE implementation of retrieve event flags

The bulk of the code of the `NUSE_Event_Group_Retrieve()` API function—after parameter checking—is selected by conditional compilation, dependent on whether support for blocking (task suspend) API calls is enabled. We will look at the two variants separately here.

If blocking is not enabled, the full code for this API call is as follows:

```
temp_events = NUSE_Event_Group_Data[group] & requested_events;

if (operation == NUSE_OR)
{
  if (temp_events != 0)
  {
    return_value = NUSE_SUCCESS;
  }
  else
```

```
  {
    return_value = NUSE_NOT_PRESENT;
  }
}
else /* operation == NUSE_AND */
{
  if (temp_events == requested_events)
  {
    return_value = NUSE_SUCCESS;
  }
  else
  {
    return_value = NUSE_NOT_PRESENT;
  }
}
```

The required event flags are extracted from the specified event flag group. This value is compared with the required events, taking into account the AND/OR operation and the result returned, along with the actual value of the requested flags.

When blocking is enabled, the code becomes more complex:

```
do
{
  temp_events = NUSE_Event_Group_Data[group] & requested_events;
  if (operation == NUSE_OR)
  {
    if (temp_events != 0)
    {
      return_value = NUSE_SUCCESS;
    }
    else
    {
      return_value = NUSE_NOT_PRESENT;
    }
  }
  else   /* operation == NUSE_AND */
  {
    if (temp_events == requested_events)
    {
      return_value = NUSE_SUCCESS;
    }
    else
    {
      return_value = NUSE_NOT_PRESENT;
    }
  }
```

```
if (return_value == NUSE_SUCCESS)
{
  suspend = NUSE_NO_SUSPEND;
}
else
{
  if (suspend == NUSE_SUSPEND)    /* block task */
  {
    NUSE_Event_Group_Blocking_Count[group]++;
    NUSE_Suspend_Task(NUSE_Task_Active, (group << 4) |
        NUSE_EVENT_SUSPEND);
    return_value =
      NUSE_Task_Blocking_Return[NUSE_Task_Active];
    if (return_value! = NUSE_SUCCESS)
    {
      suspend = NUSE_NO_SUSPEND;
    }
  }
}
} while (suspend == NUSE_SUSPEND);
```

The code is enclosed in a do...while loop, which continues while the parameter suspend has the value NUSE_SUSPEND.

The required event flags are retrieved in the same way as with a nonblocking call. If the retrieve is not successful and suspend is set to NUSE_NO_SUSPEND, the API call exits with NUSE_NOT_PRESENT. If suspend was set to NUSE_SUSPEND, the task is suspended. On return (i.e., when the task is woken up), if the return value is NUSE_SUCCESS, indicating that the task was woken because event flags in this group have been set or cleared, the code loops back to the top and the flags are retrieved and tested. Of course, since there is no reset API function for event flag groups, this is the only possible reason for the task to be woken, but the process of checking NUSE_Task_Blocking_Return[] is retained for consistency with the handling of blocking with other object types.

Event flag group utility services

Nucleus RTOS has three API calls that provide utility functions associated with event flag groups: return information about a group, return number of event flag groups in the application, and return pointers to all groups in the application. The first two of these are implemented in Nucleus SE.

Event flag group information

This service call obtains a selection of information about an event flag group. The Nucleus SE implementation differs from Nucleus RTOS in that it returns less information, as object naming and suspend ordering are not supported and task suspend may not be enabled.

Nucleus RTOS API call for event group information

Service call prototype:

```
STATUS NU_Event_Group_Information(
NU_EVENT_GROUP *group,
CHAR *name,
UNSIGNED *event_flags,
UNSIGNED *tasks_waiting,
NU_TASK **first_task);
```

Parameters:

group—pointer to the user-supplied event flag group control block;

name—pointer to an 8-character destination area for the event flag group's name; this includes space for the NULL terminator;

event_flags—a pointer to a variable, which will receive the current value of the specified event flag group;

tasks_waiting—a pointer to a variable, which will receive the number of tasks suspended on this event flag group;

first_task—a pointer to a variable of type NU_TASK, which will receive a pointer to the first suspended task.

Returns:

NU_SUCCESS—The call was completed successfully.

NU_INVALID_GROUP—The event flag group pointer is not valid.

Nucleus SE API call for event group information

This API call supports the key functionality of the Nucleus RTOS API.

Service call prototype:

```
STATUS NUSE_Event_Group_Information(
NUSE_EVENT_GROUP group,
```

```
U8 *event_flags,
U8 *tasks_waiting,
NUSE_TASK *first_task);
```

Parameters:

group—the index of the event flag group about which information is being requested;

event_flags—a pointer to a variable, which will receive the current value of the specified event flag group;

tasks_waiting—a pointer to a variable, which will receive the number of tasks suspended on this event flag group (nothing returned if task suspend is disabled);

first_task—a pointer to a variable of type NUSE_TASK, which will receive the index of the first suspended task (nothing returned if task suspend is disabled).

Returns:

NUSE_SUCCESS—The call was completed successfully.

NUSE_INVALID_GROUP—The event flag group index is not valid.

Nucleus SE implementation of event group information

The implementation of this API call is quite straightforward:

```
*event_flags = NUSE_Event_Group_Data[group];

#if NUSE_BLOCKING_ENABLE

    *tasks_waiting = NUSE_Event_Group_Blocking_Count[group];
    if (NUSE_Event_Group_Blocking_Count[group] != 0)
    {
      U8 index;

      for (index = 0; index < NUSE_TASK_NUMBER; index++)
      {
        if ((LONIB(NUSE_Task_Status[index]) ==
            NUSE_EVENT_SUSPEND)
            && (HINIB(NUSE_Task_Status[index]) == group))
        {
          *first_task = index;
          break;
        }
      }
    }
    else
    {
      *first_task = 0;
    }
```

```
#else
  *tasks_waiting = 0;
  *first_task = 0;
#endif

return NUSE_SUCCESS;
```

The function returns the value of the event flag group. Then, if blocking API calls is enabled, the number of waiting tasks and the index of the first one are returned (otherwise, these two parameters are set to 0).

Obtaining the number of event flag groups

This service call returns the number of event flag groups configured in the application. Whilst in Nucleus RTOS this will vary over time and the returned value will represent the current number of groups, in Nucleus SE the value returned is set at build time and cannot change.

Nucleus RTOS API call for event flag group count

Service call prototype:

```
UNSIGNED NU_Established_Event_Groups(VOID);
```

Parameters:

None

Returns:

The current number of created event flag groups in the application

Nucleus SE API call for event flag group count

This API call supports the key functionality of the Nucleus RTOS API.

Service call prototype:

```
U8 NUSE_Event_Group_Count(void);
```

Parameters:

None

Returns:

The number of configured event flag groups in the application

Nucleus SE implementation of event flag group count

The implementation of this API call is almost trivially simple: the value of the `#define` symbol `NUSE_EVENT_GROUP_NUMBER` is returned.

Data structures

Event flag groups utilize one or two data structures—both in RAM—which, like other Nucleus SE objects, are tables, included and dimensioned according to the number of groups configured and options selected.

I strongly recommend that application code does not access these data structures directly, but uses the provided API functions. This avoids incompatibility with future versions of Nucleus SE and unwanted side effects and simplifies porting of an application to Nucleus RTOS. The details of data structures are included here to facilitate easier understanding of the working of the service call code and for debugging.

RAM data

These data structures are:

`NUSE_Event_Group_Data[]`—This is an array of type `U8`, with one entry for each configured event flag group, and is where the event flag data are stored.

`NUSE_Event_Group_Blocking_Count[]`—This type `U8` array contains the counts of how many tasks are blocked on each event flag group. This array only exists if blocking API call support is enabled.

These data structures are all initialized to zeros by `NUSE_Init_Event_Group()` when Nucleus SE starts up. Chapter 17, *Nucleus SE Initialization* and Start-up, provides a full description of Nucleus SE start-up procedures.

Here are the definitions of these data structures in `nuse_init.c` file:

```
RAM U8 NUSE_Event_Group_Data[NUSE_EVENT_GROUP_NUMBER];

#if NUSE_BLOCKING_ENABLE

  RAM U8 NUSE_Event_Group_Blocking_Count
  [NUSE_EVENT_GROUP_NUMBER];

#endif
```

ROM data

There are no ROM data structures associated with event flag groups.

Event flag group data footprint

Like all kernel objects in Nucleus SE, the amount of data memory required for event flag groups is readily predictable.

The ROM data footprint for all the event flag groups in an application is 0.

The RAM data footprint (in bytes) for all the event flag groups in an application, when blocking API calls is enabled, is simply:

```
NUSE_EVENT_GROUP_NUMBER * 2
```

Otherwise, it is just:

```
NUSE_EVENT_GROUP_NUMBER.
```

Unimplemented API calls

Three event flag group API calls found in Nucleus RTOS are not implemented in Nucleus SE:

Create event flag group

This API call creates an event flag group. It is not needed with Nucleus SE, as event flag groups are created statically.

Service call prototype:

```
STATUS NU_Create_Event_Group(
NU_EVENT_GROUP *group,
CHAR *name);
```

Parameters:

group—pointer to a user-supplied event flag group control block; this will be used as a "handle" for the event flag group in other API calls.

name—pointers to a 7-character, NULL-terminated name for the event flag group.

Returns:

NU_SUCCESS—indicates successful completion of the service.

NU_INVALID_GROUP—indicates the event flag group control block pointer is NULL or already in use.

Delete event flag group

This API call deletes a previously created event flag group. It is not needed with Nucleus SE, as event flag groups are created statically and cannot be deleted.

Service call prototype:

```
STATUS NU_Delete_Event_Group(NU_EVENT_GROUP *group);
```

Parameters:

group—pointer to partition event flag group control block

Returns::

NU_SUCCESS—indicates successful completion of the service.

NU_INVALID_GROUP—indicates the event flag group pointer is invalid.

Event flag group pointers

This API call builds a sequential list of pointers to all event flag groups in the system. It is not needed with Nucleus SE, as event flag groups are identified by a simple index, not a pointer, and it would be redundant.

Service call prototype:

```
UNSIGNED NU_Event_Group_Pointers(
NU_EVENT_GROUP *pointer_list,
UNSIGNED maximum_pointers);
```

Parameters:

pointer_list—pointer to an array of NU_EVENT_GROUP pointers; this array will be filled with pointers to established event flag groups in the system.

maximum_pointers—the maximum number of pointers to place in the array.

Returns::

The number of NU_EVENT_GROUP pointers placed into the array.

Compatibility with Nucleus RTOS

With all aspects of Nucleus SE, it was my goal to maintain as high a level of applications code compatibility with Nucleus RTOS as possible. Event flag groups are no exception and, from a user's perspective, they are implemented in much the same way as in Nucleus RTOS. There are areas of incompatibility, which have come about where I determined that such an incompatibility would be acceptable, given that the resulting code is easier to understand, or, more likely, could be made more memory efficient. Otherwise, Nucleus RTOS API calls may be almost directly mapped onto Nucleus SE calls. Chapter 20, *Using Nucleus SE*, includes further information on using Nucleus SE for users of Nucleus RTOS.

Object identifiers

In Nucleus RTOS, all objects are described by a data structure—a control block—which has a specific data type. A pointer to this control block serves as an identifier for the event flag group. In Nucleus SE, I decided that a different approach was needed for memory efficiency, and all kernel objects are described by a number of tables in RAM and/or ROM. The size of these tables is determined by the number of each object type that is configured. The identifier for a specific object is simply an index into those tables. So, I have defined `NUSE_EVENT_GROUP` as being equivalent to `U8`; a variable—not a pointer—of this type then serves as the event flag group identifier. This is a small incompatibility, which is easily handled if code is ported to or from Nucleus RTOS. Object identifiers are normally just stored and passed around and not operated on in any way.

Nucleus RTOS also supports naming of event flag groups. These names are only used for target-based debug facilities. I omitted them from Nucleus SE to save memory.

Number of flags in a group

In Nucleus RTOS, event flag groups contain 32 flags; in Nucleus SE I reduced this to eight, as this is likely to be sufficient for simpler applications and saves RAM. Nucleus SE could be modified quite readily, if larger event flag groups were required.

Consume flags option

In Nucleus RTOS, there is an option to clear (consume) event flags when they are retrieved. I chose to omit this from Nucleus SE for simplicity, as the actual consume (deletion) of flags takes place when all the blocked tasks have received their retrieved flags; this would have been difficult to implement. The retrieving task can always clear the event flags by means of a further API call, if required.

Unimplemented API calls

Nucleus RTOS supports seven service calls to work with event flag groups. Of these, three are not implemented in Nucleus SE. Details of these and of the decision to omit them were outlined.

10

Semaphores

Semaphores were introduced in Chapter 3, *RTOS Services and Facilities*. Their primary use is the control of access to resources.

Using semaphores

In Nucleus SE, semaphores are configured at build time. There may be a maximum of 16 semaphores configured for an application. If no semaphores are configured, no data structures or service call code appertaining to semaphores is included in the application.

A semaphore is simply a counter of type **u8**, access to which is controlled so that it may be safely utilized by multiple tasks. A task can decrement (obtain) a semaphore and increment (release) it. Trying to obtain a semaphore that has the value zero may result in an error or task suspension, depending on options selected in the application program interface (API) call and the Nucleus SE configuration.

Configuring semaphores

Number of semaphores

As with most aspects of Nucleus SE, the configuration of semaphores is primarily controlled by **#define** statements in **nuse_config.h**. The key setting is **NUSE_SEMAPHORE_NUMBER**, which determines how many semaphores are configured for the application. The default setting is 0 (i.e., no semaphores are in use) and you can set it to any value up to 16. An erroneous value will result in a compile time error, which is generated by a test in **nuse_config_check.h** (this is included into **nuse_config.c** and hence compiled with this module), resulting in a **#error** statement being compiled.

Choosing a nonzero value is the "master enable" for semaphores. This results in some data structures being defined and sized accordingly, of which more later in this chapter. It also activates the API enabling settings.

Embedded RTOS Design. DOI: https://doi.org/10.1016/B978-0-12-822851-7.00010-2

API enables

Every API function (service call) in Nucleus SE has an enabling #define symbol in nuse_config.h. For semaphores, these are:

```
NUSE_SEMAPHORE_OBTAIN
NUSE_SEMAPHORE_RELEASE
NUSE_SEMAPHORE_RESET
NUSE_SEMAPHORE_INFORMATION
NUSE_SEMAPHORE_COUNT
```

By default, all of these are set to FALSE, thus disabling each service call and inhibiting the inclusion of any implementation code. To configure semaphores for an application, you need to select the API calls that you want to use and set their enabling symbols to TRUE.

Here is an extract from the default nuse_config.h file.

```
#define NUSE_SEMAPHORE_NUMBER   0   /* Number of semaphores in
                        the system - 0-16 */

#define NUSE_SEMAPHORE_OBTAIN   FALSE   /* Service call enabler */
#define NUSE_SEMAPHORE_RELEASE  FALSE   /* Service call enabler */
#define NUSE_SEMAPHORE_RESET    FALSE  /* Service call enabler */
#define NUSE_SEMAPHORE_INFORMATION  FALSE   /* Service call
                                               enabler */
#define NUSE_SEMAPHORE_COUNT    FALSE  /* Service call enabler */
```

A compile time error will result if a semaphore API function is enabled and no semaphores are configured (except for NUSE_Semaphore_Count(), which is always permitted). If your code uses an API call, which has not been enabled, a link time error will result, as no implementation code will have been included in the application.

Semaphore service calls

Nucleus RTOS supports eight service calls that appertain to semaphores, which provide the following functionality:

Obtain a semaphore. Implemented by NUSE_Semaphore_Obtain() in Nucleus SE.

Release a semaphore. Implemented by NUSE_Semaphore_Release() in Nucleus SE.

Restore a semaphore to the unused state, with no tasks suspended (reset). Implemented by NUSE_Semaphore_Reset() in Nucleus SE.

Provide information about a specified semaphore. Implemented by `NUSE_Semaphore_Information()` in Nucleus SE.

Return a count of how many semaphores are (currently) configured for the application. Implemented by `NUSE_Semaphore_Count()` in Nucleus SE.

Add a new semaphore to the application (create). Not implemented in Nucleus SE.

Remove a semaphore from the application (delete). Not implemented in Nucleus SE.

Return pointers to all the semaphores (currently) in the application. Not implemented in Nucleus SE.

The implementation of each of these service calls will be examined in detail.

Semaphore obtain and release services

The fundamental operations, which can be performed on a semaphore, are obtaining (decrementing) and releasing (incrementing) it. Nucleus RTOS and Nucleus SE each provide two basic API calls for these operations, which will be discussed here.

Obtaining a semaphore

The Nucleus RTOS API call for obtaining a semaphore is very flexible, enabling you to suspend indefinitely, or with a timeout, if the operation cannot be completed immediately; that is, you try to obtain a semaphore that currently has the value zero. Nucleus SE provides the same service, except task suspend is optional and timeout is not implemented.

Nucleus RTOS API call for obtaining a semaphore

Service call prototype:

```
STATUS NU_Obtain_Semaphore(
NU_SEMAPHORE *semaphore,
UNSIGNED suspend);
```

Parameters:

semaphore—pointer to the user-supplied semaphore control block;

suspend—specification for task suspend; may be `NU_NO_SUSPEND` or `NU_SUSPEND` or a timeout value.

Returns:

NU_SUCCESS—The call was completed successfully.

NU_UNAVAILABLE—The semaphore had the value zero.

NU_INVALID_SEMAPHORE—The semaphore pointer is invalid.

NU_INVALID_SUSPEND—Suspend was attempted from a nontask thread.

NU_SEMAPHORE_WAS_RESET—The semaphore was reset, while the task was suspended.

Nucleus SE API Call for obtaining a semaphore

This API call supports the key functionality of the Nucleus RTOS API.

Service call prototype:

```
STATUS NUSE_Semaphore_Obtain(
NUSE_SEMAPHORE semaphore,
U8 suspend);
```

Parameters:

semaphore—the index (ID) of the semaphore to be utilized;

suspend—specification for task suspend; may be **NUSE_NO_SUSPEND** or **NUSE_SUSPEND**.

Returns:

NUSE_SUCCESS—The call was completed successfully.

NUSE_UNAVAILABLE—The semaphore had the value zero.

NUSE_INVALID_SEMAPHORE—The semaphore index is invalid.

NUSE_INVALID_SUSPEND—Suspend was attempted from a nontask thread or when blocking API calls were not enabled.

NUSE_SEMAPHORE_WAS_RESET—The semaphore was reset, while the task was suspended.

Nucleus SE implementation of obtain semaphore

The bulk of the code of the **NUSE_Semaphore_Obtain()** API function—after parameter checking—is selected by conditional compilation, dependent on whether support for blocking (task suspend) API calls is enabled. We will look at the two variants separately here.

If blocking is not enabled, the logic for this API call is quite simple:

```
if (NUSE_Semaphore_Counter[semaphore] != 0) /* semaphore
                                             available */
{
  NUSE_Semaphore_Counter[semaphore]--;
  return_value = NUSE_SUCCESS;
}
else                    /* semaphore unavailable */
{
  return_value = NUSE_UNAVAILABLE;
}
```

The semaphore value is tested and, if nonzero, decremented. When blocking is enabled, the logic becomes more complex:

```
do
{
  if (NUSE_Semaphore_Counter[semaphore] != 0) /* semaphore
                                               available */
  {
    NUSE_Semaphore_Counter[semaphore]--;
    return_value = NUSE_SUCCESS;
    suspend = NUSE_NO_SUSPEND;
  }
  else                        /* semaphore unavailable */
  {
    if (suspend == NUSE_NO_SUSPEND)
    {
      return_value = NUSE_UNAVAILABLE;
    }
    else
    {                     /* block task */
      NUSE_Semaphore_Blocking_Count[semaphore]++;
      NUSE_Suspend_Task(NUSE_Task_Active,
            semaphore << 4) | NUSE_SEMAPHORE_SUSPEND);
      return_value =
      NUSE_Task_Blocking_Return[NUSE_Task_Active];
      if (return_value! = NUSE_SUCCESS)
      {
        suspend = NUSE_NO_SUSPEND;
      }
    }
  }
} while (suspend == NUSE_SUSPEND);
```

Some explanation of the code may be useful:

The code is enclosed in a `do...while` loop, which continues while the parameter suspend has the value `NUSE_SUSPEND`.

If the semaphore is nonzero, it is decremented. The `suspend` variable is set to `NUSE_NO_SUSPEND` and the API call exits with `NUSE_SUCCESS`.

If the semaphore is zero and `suspend` is set to `NUSE_NO_SUSPEND`, the API call exits with `NUSE_UNAVAILABLE`. If suspend was set to `NUSE_SUSPEND`, the task is suspended. On return (i.e., when the task is woken up), if the return value is `NUSE_SUCCESS`, indicating that the task was woken because the semaphore has been released (as opposed to a reset of the semaphore), the code loops back to the top.

Releasing a semaphore

The Nucleus RTOS API call for releasing a semaphore is quite simple; the semaphore is incremented and success reported. Nucleus SE provides the same service, except an overflow check is performed.

Nucleus RTOS API call for releasing a semaphore

Service call prototype:

```
STATUS NU_Release_Semaphore(NU_SEMAPHORE *semaphore);
```

Parameters:

`semaphore`—pointer to the user-supplied semaphore control block.

Returns:

`NU_SUCCESS`—The call was completed successfully.

`NU_INVALID_SEMAPHORE`—The semaphore pointer is invalid.

Nucleus SE API call for releasing a semaphore

This API call supports the key functionality of the Nucleus RTOS API.

Service call prototype:

```
STATUS NUSE_Semaphore_Release(NUSE_SEMAPHORE semaphore);
```

Parameters:

semaphore—the index (ID) of the semaphore to be released.

Returns:

NUSE_SUCCESS—The call was completed successfully.

NUSE_INVALID_SEMAPHORE—The semaphore index is invalid.

NUSE_UNAVAILABLE—The semaphore has the value 255 and cannot be incremented.

Nucleus SE implementation of release semaphore

The initial code of the NUSE_Semaphore_Release() API function—after parameter checking—is common, whether task blocking is enabled or not. The value of the semaphore is checked and, if it is less than 255, incremented.

Further code is selected by conditional compilation, if support for blocking (task suspend) API calls is enabled:

```
NUSE_CS_Enter();

if (NUSE_Semaphore_Counter[semaphore] < 255)
{
    NUSE_Semaphore_Counter[semaphore]++;
    return_value = NUSE_SUCCESS;

    #if NUSE_BLOCKING_ENABLE
        if (NUSE_Semaphore_Blocking_Count[semaphore] != 0)
        {
            U8 index;      /* check whether a task is blocked */
                           /* on this semaphore */

            NUSE_Semaphore_Blocking_Count[semaphore]--;
            for (index = 0; index < NUSE_TASK_NUMBER; index++)
            {
              if ((LONIB(NUSE_Task_Status[index]) ==
                  NUSE_SEMAPHORE_SUSPEND)
                  && (HINIB(NUSE_Task_Status[index]) ==
                  semaphore))
                {
                  NUSE_Task_Blocking_Return[index] =
                    NUSE_SUCCESS;
                  NUSE_Wake_Task(index);
                  break;
                }
            }
        }

    #endif
}
```

```
    else
    {
       return_value = NUSE_UNAVAILABLE;
    }

    NUSE_CS_Exit();

    return return_value;
```

If any tasks are suspended on this semaphore, the first one is woken up.

Semaphore utility services

Nucleus RTOS has four API calls that provide utility functions associated with semaphores: reset semaphore, return information about a semaphore, return number of semaphores in the application, and return pointers to all semaphores in the application. The first three of these are implemented in Nucleus SE.

Resetting a semaphore

This API call restores the semaphore to its initial, unused state. This API function is unusual, compared with that available for other kernel objects, as, although it is a reset, it does not simply initialize its counter to the start-up value; a new initial count is provided in the call. Any tasks that were suspended on the semaphore are resumed and receive a return code of **NUSE_SEMAPHORE_WAS_RESET** (in Nucleus SE, or **NU_SEMAPHORE_RESET** with Nucleus RTOS).

Nucleus RTOS API call for resetting a semaphore

Service call prototype:

```
STATUS NU_Reset_Semaphore(
NU_SEMAPHORE *semaphore,
UNSIGNED initial_count);
```

Parameters:

semaphore—pointer to user-supplied semaphore control block;

initial_count—the value to which the semaphore's counter is to be set.

Returns:

NU_SUCCESS—The call was completed successfully.

NU_INVALID_SEMAPHORE—The semaphore pointer is not valid.

Nucleus SE API call for resetting a semaphore

This API call supports the key functionality of the Nucleus RTOS API.

Service call prototype:

```
STATUS NUSE_Semaphore_Reset (
NUSE_SEMAPHORE semaphore,
U8 initial_count);
```

Parameters:

semaphore—the index (ID) of the semaphore to be reset;

initial_count—the value to which the semaphore's counter is to be set.

Returns:

NUSE_SUCCESS—The call was completed successfully.

NUSE_INVALID_SEMAPHORE—The semaphore index is not valid.

Nucleus SE implementation of semaphore reset

The main job of the NUSE_Semaphore_Reset() API function—after parameter checking—is to simply set the appropriate entry in NUSE_Semaphore_Counter[] to the provided initial value.

When blocking is enabled, further code is required to unblock tasks:

```
while (NUSE_Semaphore_Blocking_Count [semaphore] != 0)
{
  U8 index;      /* check whether any tasks are blocked */
           /* on this semaphore */
  for (index = 0; index < NUSE_TASK_NUMBER; index++)
  {
    if ((LONIB(NUSE_Task_Status [index]) ==
        NUSE_SEMAPHORE_SUSPEND)
        && (HINIB(NUSE_Task_Status [index]) == semaphore))
    {
      NUSE_Task_Blocking_Return [index] =
      NUSE_SEMAPHORE_WAS_RESET;
      NUSE_Task_Status [index] = NUSE_READY;
      break;
    }
  }
  NUSE_Semaphore_Blocking_Count [semaphore] --;
}

#if NUSE_SCHEDULER_TYPE == NUSE_PRIORITY_SCHEDULER
  NUSE_Reschedule(NUSE_NO_TASK);
#endif
```

Each task suspended on the semaphore is marked as *Ready* with a suspend return code of NUSE_SEMAPHORE_WAS_RESET. After this process is complete, if the Priority scheduler is in use, a call is made to NUSE_Reschedule(), as one or more higher priority tasks may have been readied and needs to be allowed to run.

Semaphore information

This service call obtains a selection of information about a semaphore. The Nucleus SE implementation differs from Nucleus RTOS in that it returns less information, as object naming and suspend ordering are not supported and task suspend may not be enabled.

Nucleus RTOS API call for semaphore information

Service call prototype:

```
STATUS NU_Semaphore_Information(
NU_SEMAPHORE *semaphore,
CHAR *name,
UNSIGNED *current_count,
OPTION *suspend_type,
UNSIGNED *tasks_waiting,
NU_TASK **first_task);
```

Parameters:

semaphore—pointer top the control block of the semaphore about which information is being requested;

name—pointer to an 8-character destination area for the semaphore's name; this includes space for the NULL terminator;

current_count—a pointer to a variable, which will receive the current value of the semaphore counter;

suspend_type—pointer to a variable that holds the task's suspend type; valid task suspend types are NU_FIFO and NU_PRIORITY;

tasks_waiting—a pointer to a variable, which will receive the number of tasks suspended on this semaphore;

first_task—a pointer to a variable of type NU_TASK, which will receive a pointer to the control block of the first suspended task.

Returns:

NU_SUCCESS—The call was completed successfully.

NU_INVALID_SEMAPHORE—The semaphore pointer is not valid.

Nucleus SE API call for semaphore information

This API call supports the key functionality of the Nucleus RTOS API.

Service call prototype:

```
STATUS NUSE_Semaphore_Information(
NUSE_SEMAPHORE semaphore,
U8 *current_count,
U8 *tasks_waiting,
NUSE_TASK *first_task);
```

Parameters:

semaphore—the index of the semaphore about which information is being requested;

current_count—a pointer to a variable, which will receive the current value of the semaphore counter;

tasks_waiting—a pointer to a variable, which will receive the number of tasks suspended on this semaphore (nothing returned if task suspend is disabled);

first_task—a pointer to a variable of type NUSE_TASK, which will receive the index of the first suspended task (nothing returned if task suspend is disabled).

Returns:

NUSE_SUCCESS—The call was completed successfully.

NUSE_INVALID_SEMAPHORE—The semaphore index is not valid.

NUSE_INVALID_POINTER—One or more of the pointer parameters is invalid.

Nucleus SE implementation of semaphore information

The implementation of this API call is quite straightforward:

```
NUSE_CS_Enter();

*current_count = NUSE_Semaphore_Counter[semaphore];
#if NUSE_BLOCKING_ENABLE

  *tasks_waiting = NUSE_Semaphore_Blocking_Count[semaphore];
  if (NUSE_Semaphore_Blocking_Count[semaphore] != 0)
  {
    U8 index;

    for (index = 0; index < NUSE_TASK_NUMBER; index++)
```

```
        {
           if ((LONIB(NUSE_Task_Status[index]) ==
                NUSE_SEMAPHORE_SUSPEND) &&
                (HINIB(NUSE_Task_Status[index]) == semaphore))
           {
                *first_task = index;
                break;
           }
        }
     }
     else
     {
        *first_task = 0;
     }

#else

     *tasks_waiting = 0;
     *first_task = 0;

#endif

NUSE_CS_Exit();

return NUSE_SUCCESS;
```

The function returns the semaphore status. Then, if blocking API calls is enabled, the number of waiting tasks and the index of the first one are returned (otherwise, these two parameters are set to 0).

Obtaining the number of semaphores

This service call returns the number of semaphores configured in the application. Whilst in Nucleus RTOS this will vary over time and the returned value will represent the current number of semaphores, in Nucleus SE the value returned is set at build time and cannot change.

Nucleus RTOS API call for semaphore count

Service call prototype:

```
UNSIGNED NU_Established_Semaphores(VOID);
```

Parameters:

None

Returns:

The number of created semaphores in the application

Nucleus SE API call for semaphore count

This API call supports the key functionality of the Nucleus RTOS API.

Service call prototype:

```
U8 NUSE_Semaphore_Count(void);
```

Parameters:

None

Returns:

The number of configured semaphores in the application

Nucleus SE implementation of semaphore count

The implementation of this API call is almost trivially simple: the value of the #define symbol NUSE_SEMAPHORE_NUMBER is returned.

Data structures

Semaphores utilize two or three data structures—in RAM and ROM—which, like other Nucleus SE objects, are a series of tables, included and dimensioned according to the number of semaphores configured and options selected.

I strongly recommend that application code does not access these data structures directly but uses the provided API functions. This avoids incompatibility with future versions of Nucleus SE and unwanted side effects and simplifies porting of an application to Nucleus RTOS. The details of data structures are included here to facilitate easier understanding of the working of the service call code and for debugging.

RAM data

These data structures are:

NUSE_Semaphore_Counter[]—This is an array of type U8, with one entry for each configured semaphore; this is where the count value is stored.

NUSE_Semaphore_Blocking_Count[]—This type U8 array contains the counts of how many tasks are blocked on each semaphore. This array only exists if blocking API call support is enabled.

NUSE_Semaphore_Counter[] is initialized to an initial value (see "ROM data" section) and NUSE_Semaphore_Blocking_Count[]

is set to zero by `NUSE_Init_Semaphore()` when Nucleus SE starts up. Chapter 17, *Nucleus SE Initialization* and Start-up, provides a full description of Nucleus SE start-up procedures.

Here are the definitions of these data structures in `nuse_init.c` file.

```
RAM U8 NUSE_Semaphore_Counter[NUSE_SEMAPHORE_NUMBER];
#if NUSE_BLOCKING_ENABLE

   RAM U8 NUSE_Semaphore_Blocking_Count[NUSE_SEMAPHORE_NUMBER];

#endif
```

ROM data

This data structure is:

`NUSE_Semaphore_Initial_Value[]`—This is an array of type `U8`, with one entry for each configured semaphore; this is the value to which the counters are initialized.

This data structure is declared and initialized (statically, of course) in `nuse_config.c`:

```
ROM U8 NUSE_Semaphore_Initial_Value[NUSE_SEMAPHORE_NUMBER] =
{
  /* semaphore initial count values */
};
```

Semaphore data footprint

Like all kernel objects in Nucleus SE, the amount of data memory required for semaphores is readily predictable.

The ROM data footprint (in bytes) for all the semaphores in an application is just:

`NUSE_SEMAPHORE_NUMBER`

The RAM data footprint (in bytes) for all the semaphores in an application, when blocking API calls is enabled, may be computed thus:

`NUSE_SEMAPHORE_NUMBER * 2`

Otherwise, it is just `NUSE_SEMAPHORE_NUMBER`.

Unimplemented API calls

Three semaphore API calls found in Nucleus RTOS are not implemented in Nucleus SE:

Create semaphore

This API call creates a semaphore. It is not needed with Nucleus SE, as semaphores are created statically.

Service call prototype:

```
STATUS NU_Create_Semaphore(
NU_SEMAPHORE *semaphore,
CHAR *name,
UNSIGNED initial_count,
OPTION suspend_type);
```

Parameters:

semaphore—pointer to a user-supplied semaphore control block; this will be used as a "handle" for the semaphore in other API calls;

name—pointers to a 7-character, NULL-terminated name for the semaphore;

initial_count—the initial value of the semaphore;

suspend_type—specifies how tasks suspend on the semaphore. Valid options for this parameter are NU_FIFO and NU_PRIORITY, which represent first-in first-out (FIFO) and priority-order task suspension, respectively.

Returns:

NU_SUCCESS—indicates successful completion of the service.

NU_INVALID_SEMAPHORE—indicates the semaphore control block pointer is NULL.

NU_INVALID_SUSPEND—indicates that the suspend_type parameter is invalid.t

Delete semaphore

This API call deletes a previously created semaphore. It is not needed with Nucleus SE, as semaphores are created statically and cannot be deleted.
Service call prototype:

```
STATUS NU_Delete_Semaphore(NU_SEMAPHORE *semaphore);
```

Parameters:

semaphore—pointer to semaphore control block

Returns:

NU_SUCCESS—indicates successful completion of the service.

NU_INVALID_SEMAPHORE—indicates the semaphore pointer is invalid.

Semaphore pointers

This API call builds a sequential list of pointers to all semaphores in the system. It is not needed with Nucleus SE, as semaphores are identified by a simple index, not a pointer, and it would be redundant.

Service call prototype:

```
UNSIGNED NU_Semaphore_Pointers(
NU_SEMAPHORE **pointer_list,
UNSIGNED maximum_pointers);
```

Parameters:

pointer_list—pointer to an array of NU_SEMAPHORE pointers; this array will be filled with pointers to established semaphores in the system;

maximum_pointers—the maximum number of pointers to place in the array.

Returns:

The number of NU_SEMAPHORE pointers placed into the array

Compatibility with Nucleus RTOS

With all aspects of Nucleus SE, it was my goal to maintain as high a level of applications code compatibility with Nucleus RTOS as possible. Semaphores are no exception, and, from a user's perspective, they are implemented in much the same way as in Nucleus RTOS. There are areas of incompatibility, which have come about where I determined that such an incompatibility would be acceptable, given that the resulting code is easier to understand, or, more likely, could be made more memory efficient. Otherwise, Nucleus RTOS API calls may be almost directly mapped onto Nucleus SE calls. Chapter 20, *Using Nucleus SE*, includes further information on using Nucleus SE for users of Nucleus RTOS.

Object identifiers

In Nucleus RTOS, all objects are described by a data structure—a control block—which has a specific data type. A pointer to this control block serves as an identifier for the semaphore. In Nucleus SE, I decided that a different approach was needed for memory efficiency, and all kernel objects are described by a number of tables in RAM and/or ROM. The size of these tables is determined by the number of each object type that is configured. The identifier for a specific object is simply an index into those tables. So, I have defined NUSE_SEMAPHORE as being equivalent to u8; a variable—not a pointer—of this type then serves as the semaphore identifier. This is a small incompatibility, which is easily handled if code is ported to or from Nucleus RTOS. Object identifiers are normally just stored and passed around and not operated on in any way.

Nucleus RTOS also supports naming of semaphores. These names are only used for target-based debug facilities. I omitted them from Nucleus SE to save memory.

Counter size

In Nucleus RTOS, a semaphore counter is an unsigned, which is generally a 32-bit variable. In Nucleus SE, the counter is 8 bits; this could be easily changed. Normally, there is no check on overflow for semaphore releases in Nucleus RTOS. The Nucleus SE API call will not allow the counter to be incremented beyond 255.

Unimplemented API calls

Nucleus RTOS supports eight service calls to work with semaphores. Of these, three are not implemented in Nucleus SE. Details of these and of the decision to omit them were outlined earlier in this chapter.

11

Mailboxes

Mailboxes were introduced in Chapter 3, *RTOS Services and Facilities*. They are perhaps the second simplest method of intertask communication—after signals—supported by Nucleus SE. They provide a low-cost, but flexible, means of passing simple messages between tasks.

Using mailboxes

In Nucleus SE, mailboxes are configured at build time. There may be a maximum of 16 mailboxes configured for an application. If no mailboxes are configured, no data structures or service call code appertaining to mailboxes are included in the application.

A mailbox is simply a storage location, big enough to hold a single variable of type **ADDR**, access to which is controlled so that it may be safely utilized by multiple tasks. One task can write to a mailbox. It is then full, and no task can write to it until a task does a read on the mailbox or the mailbox is reset. Trying to write to a full mailbox or read from an empty one may result in an error or task suspension, depending on options selected in the application program interface (API) call and the Nucleus SE configuration.

Mailboxes and queues

In some operating system implementations, mailboxes are not supported, and the use of a single-entry queue is recommended as an alternative. This sounds reasonable, as such a queue would provide the same functionality as a mailbox. However, a queue is a rather more complex data structure than a mailbox and carries considerably more overhead in data (head and tail pointers, etc.), code and execution time.

With Nucleus SE, like Nucleus RTOS, you have the choice of both object types and can make the decision for yourself.

Embedded RTOS Design. DOI: https://doi.org/10.1016/B978-0-12-822851-7.00011-4

It is, however, worth considering this alternative approach if your application includes multiple queues, but perhaps a single mailbox. Replacing that mailbox with a queue will incur a small data overhead, but eliminates all the mailbox-related API code. It would be very easy to configure the application both ways and compare the memory footprints and performance.

Queues will be discussed in the next chapter.

Configuring mailboxes

Number of mailboxes

As with most aspects of Nucleus SE, the configuration of mailboxes is primarily controlled by `#define` statements in `nuse_config.h`. The key setting is `NUSE_MAILBOX_NUMBER`, which determines how many mailboxes are configured for the application. The default setting is 0 (i.e., no mailboxes are in use) and you can set it to any value up to 16. An erroneous value will result in a compile time error, which is generated by a test in `nuse_config_check.h` (this is included into `nuse_config.c` and hence compiled with this module), resulting in a `#error` statement being compiled.

Choosing a nonzero value is the "master enable" for mailboxes. This results in some data structures being defined and sized accordingly, of which more later in this chapter. It also activates the API enabling settings.

API enables

Every API function (service call) in Nucleus SE has an enabling `#define` symbol in `nuse_config.h`. For mailboxes, these are:

```
NUSE_MAILBOX_SEND
NUSE_MAILBOX_RECEIVE
NUSE_MAILBOX_RESET
NUSE_MAILBOX_INFORMATION
NUSE_MAILBOX_COUNT
```

By default, all of these are set to `FALSE`, thus disabling each service call and inhibiting the inclusion of any implementation code. To configure mailboxes for an application, you need to select the API calls that you want to use and set their enabling symbols to `TRUE`.

Here is an extract from the default `nuse_config.h` file.

```
#define NUSE_MAILBOX_NUMBER  0
                    /* Number of mailboxes in the system - 0-16 */

                    /* Service call enablers: */
#define NUSE_MAILBOX_SEND    FALSE
#define NUSE_MAILBOX_RECEIVE    FALSE
#define NUSE_MAILBOX_RESET    FALSE
#define NUSE_MAILBOX_INFORMATION    FALSE
#define NUSE_MAILBOX_COUNT    FALSE
```

A compile time error will result if a mailbox API function is enabled and no mailboxes are configured (except for `NUSE_Mailbox_Count()`, which is always permitted). If your code uses an API call, which has not been enabled, a link time error will result, as no implementation code will have been included in the application.

Mailbox service calls

Nucleus RTOS supports nine service calls that appertain to mailboxes, which provide the following functionality:

Send a message to a mailbox. Implemented by `NUSE_Mailbox_Send()` in Nucleus SE.

Receive a message from a mailbox. Implemented by `NUSE_Mailbox_Receive()` in Nucleus SE.

Restore a mailbox to the unused state, with no tasks suspended (reset). Implemented by `NUSE_Mailbox_Reset()` in Nucleus SE.

Provide information about a specified mailbox. Implemented by `NUSE_Mailbox_Information()` in Nucleus SE.

Return a count of how many mailboxes are (currently) configured for the application. Implemented by `NUSE_Mailbox_Count()` in Nucleus SE.

Add a new mailbox to the application (create). Not implemented in Nucleus SE.

Remove a mailbox from the application (delete). Not implemented in Nucleus SE.

Return pointers to all the mailboxes (currently) in the application. Not implemented in Nucleus SE.

Send a message to all the tasks that are suspended on a mailbox (broadcast). Not implemented in Nucleus SE.

The implementation of each of these service calls is examined in detail.

Mailbox write and read services

The fundamental operations, which can be performed on a mailbox, are writing data to it—which is sometimes termed *sending* or *posting*—and reading data from it—which is also termed *receiving*. Nucleus RTOS and Nucleus SE each provide two basic API calls for these operations, which will be discussed here.

Writing to a mailbox

The Nucleus RTOS API call for writing to a mailbox is very flexible, enabling you to suspend indefinitely, or with a timeout, if the operation cannot be completed immediately; that is, you try to write to a full mailbox. Nucleus SE provides the same service, except task suspend is optional and timeout is not implemented.

Nucleus RTOS also offers a facility to broadcast to a mailbox, but this is not supported by Nucleus SE. It will be described under the "*Unimplemented API calls*" section later in this chapter.

Nucleus RTOS API call for sending to a mailbox

Service call prototype:

```
STATUS NU_Send_To_Mailbox(
NU_MAILBOX *mailbox,
VOID *message,
UNSIGNED suspend);
```

Parameters:

`mailbox`—pointer to the mailbox to be utilized;

`message`—a pointer to the message to be sent, which is four unsigned elements;

`suspend`—specification for task suspend; may be `NU_NO_SUSPEND` or `NU_SUSPEND` or a timeout value.

Returns:

`NU_SUCCESS`—The call was completed successfully.

`NU_INVALID_MAILBOX`—The mailbox pointer is invalid.

`NU_INVALID_POINTER`—The message pointer is `NULL`.

NU_INVALID_SUSPEND—Suspend was attempted from a non-task thread.

NU_MAILBOX_FULL—The mailbox is full and suspend was not specified.

NU_TIMEOUT—The mailbox is still full even after suspending for the specified period.

NU_MAILBOX_DELETED—The mailbox was deleted, while the task was suspended.

NU_MAILBOX_WAS_RESET—The mailbox was reset, while the task was suspended.

Nucleus SE API call for sending to a mailbox

This API call supports the key functionality of the Nucleus RTOS API.

Service call prototype:

```
STATUS NUSE_Mailbox_Send(
NUSE_MAILBOX mailbox,
ADDR *message,
U8 suspend);
```

Parameters:

mailbox—the index (ID) of the mailbox to be utilized;

message—a pointer to the message to be sent, which is a single variable of type ADDR;

suspend—specification for task suspend; may be NUSE_NO_SUSPEND or NUSE_SUSPEND.

Returns:

NUSE_SUCCESS—The call was completed successfully.

NUSE_INVALID_MAILBOX—The mailbox index is invalid.

NUSE_INVALID_POINTER—The message pointer is NULL.

NUSE_INVALID_SUSPEND—Suspend was attempted from a non-task thread or when blocking API calls were not enabled.

NUSE_MAILBOX_FULL—The mailbox is full and suspend was not specified.

NUSE_MAILBOX_WAS_RESET—The mailbox was reset, while the task was suspended.

Nucleus SE implementation of mailbox send

The bulk of the code of the `NUSE_Mailbox_Send()` API function—after parameter checking—is selected by conditional compilation, dependent on whether support for blocking (task suspend) API calls is enabled. We will look at the two variants separately here.

If blocking is not enabled, the logic for this API call is quite simple and the code requires little explanation:

```
if (NUSE_Mailbox_Status[mailbox])      /* mailbox full */
{
  return_value = NUSE_MAILBOX_FULL;
}
else                      /* mailbox empty */
{
  NUSE_Mailbox_Data[mailbox] = *message;
  NUSE_Mailbox_Status[mailbox] = TRUE;
  return_value = NUSE_SUCCESS;
}
```

The message is stored in the appropriate element of `NUSE_Mailbox_Data[]` and the mailbox marked as being in use.

When blocking is enabled, the code becomes more complex:

```
do
{
  if (!NUSE_Mailbox_Status[mailbox])      /* mailbox empty */
  {
    NUSE_Mailbox_Data[mailbox] = *message;
    NUSE_Mailbox_Status[mailbox] = TRUE;
    if (NUSE_Mailbox_Blocking_Count[mailbox] != 0)
    {
      U8 index;      /* check whether a task is blocked */
            /* on this mailbox */
      NUSE_Mailbox_Blocking_Count[mailbox]--;
      for (index = 0; index < NUSE_TASK_NUMBER; index++)
      {
        if ((LONIB(NUSE_Task_Status[index]) ==
                                  NUSE_MAILBOX_SUSPEND)
          && (HINIB(NUSE_Task_Status[index]) == mailbox))
        {
          NUSE_Task_Blocking_Return[index] = NUSE_SUCCESS;
          NUSE_Wake_Task(index);
          break;
        }
      }
    }
  }
```

```
   return_value = NUSE_SUCCESS;
   suspend = NUSE_NO_SUSPEND;
 }
 else            /* mailbox full */
 {
   if (suspend == NUSE_NO_SUSPEND)
   {
     return_value = NUSE_MAILBOX_FULL;
   }
   else
   {                 /* block task */
     NUSE_Mailbox_Blocking_Count[mailbox]++;
     NUSE_Suspend_Task(NUSE_Task_Active, (mailbox << 4) |
                                   NUSE_MAILBOX_SUSPEND);
     return_value = NUSE_Task_Blocking_Return
       [NUSE_Task_Active];
     if (return_value! = NUSE_SUCCESS)
     {
       suspend = NUSE_NO_SUSPEND;
     }
   }
 }
} while (suspend == NUSE_SUSPEND);
```

Some explanation may be useful:

The code is enclosed in a do...while loop, which continues while the parameter suspend has the value NUSE_SUSPEND.

If the mailbox is empty, the supplied message is stored, and the mailbox status changed to indicate that it is full. A check is made on whether any tasks are suspended (waiting to receive) on the mailbox. If there are any tasks waiting, the first one is woken. The suspend variable is set to NUSE_NO_SUSPEND and the API call exits with NUSE_SUCCESS.

If the mailbox is full and suspend is set to NUSE_NO_SUSPEND, the API call exits with NUSE_MAILBOX_FULL. If suspend was set to NUSE_SUSPEND, the task is suspended. On return (i.e., when the task is woken up), if the return value is NUSE_SUCCESS, indicating that the task was woken because a message had been read (as opposed to a reset of the mailbox) the code loops back to the top.

Reading from a mailbox

The Nucleus RTOS API call for reading from a mailbox is very flexible, enabling you to suspend indefinitely, or with a timeout, if the operation cannot be completed immediately; that is, you try to read from an empty mailbox. Nucleus SE provides the same service, except task suspend is optional and timeout is not implemented.

Nucleus RTOS API call for receiving from a mailbox

Service call prototype:

```
STATUS NU_Receive_From_Mailbox(
NU_MAILBOX *mailbox,
VOID *message,
UNSIGNED suspend);
```

Parameters:

mailbox—pointer to the user-supplied mailbox control block;

message—a pointer to storage for the message to be received, which is the size of four unsigned variables;

suspend—specification for task suspend; may be NUSE_NO_SUSPEND or NUSE_SUSPEND or a timeout value.

Returns:

NU_SUCCESS—The call was completed successfully.

NU_INVALID_MAILBOX—The mailbox pointer is invalid.

NU_INVALID_POINTER—The message pointer is NULL.

NU_INVALID_SUSPEND—Suspend was attempted from a non-task thread.

NU_MAILBOX_EMPTY—The mailbox is empty and suspend was not specified.

NU_TIMEOUT—indicates that the mailbox is still empty even after suspending for the specified timeout value.

NU_MAILBOX_DELETED—The mailbox was deleted, while the task was suspended.

NU_MAILBOX_WAS_RESET—the mailbox was reset, while the task was suspended

Nucleus SE API call for receiving from a mailbox

This API call supports the key functionality of the Nucleus RTOS API.

Service call prototype:

```
STATUS NUSE_Mailbox_Receive(
NUSE_MAILBOX mailbox,
```

```
ADDR *message,
U8 suspend);
```

Parameters:

`mailbox`—the index (ID) of the mailbox to be utilized;

`message`—a pointer to storage for the message to be received, which is a single variable of type `ADDR`;

`suspend`—specification for task suspend; may be `NUSE_NO_SUSPEND` or `NUSE_SUSPEND`.

Returns:

`NUSE_SUCCESS`—The call was completed successfully.

`NUSE_INVALID_MAILBOX`—The mailbox index is invalid.

`NUSE_INVALID_POINTER`—The message pointer is `NULL`.

`NUSE_INVALID_SUSPEND`—Suspend was attempted from a non-task thread or when blocking API calls were not enabled.

`NUSE_MAILBOX_EMPTY`—The mailbox is empty and suspend was not specified.

`NUSE_MAILBOX_WAS_RESET`—The mailbox was reset, while the task was suspended.

Nucleus SE implementation of mailbox receive

The bulk of the code of the `NUSE_Mailbox_Receive()` API function—after parameter checking—is selected by conditional compilation, dependent on whether support for blocking (task suspend) API calls is enabled. We will look at the two variants separately here.

If blocking is not enabled, the logic for this API call is quite simple and the code requires little explanation:

```
if (!NUSE_Mailbox_Status[mailbox])      /* mailbox empty */
{
  return_value = NUSE_MAILBOX_EMPTY;
}
else
{                               /* mailbox full */
  *message = NUSE_Mailbox_Data[mailbox];
  NUSE_Mailbox_Status[mailbox] = FALSE;
  return_value = NUSE_SUCCESS;
}
```

The message is extracted from the appropriate element of `NUSE_Mailbox_Data[]` and the mailbox marked as being empty.

When blocking is enabled, the code becomes more complex:

```
do
{
  if (NUSE_Mailbox_Status[mailbox])     /* mailbox full */
  {
    *message = NUSE_Mailbox_Data[mailbox];
    NUSE_Mailbox_Status[mailbox] = FALSE;
    if (NUSE_Mailbox_Blocking_Count[mailbox] ! = 0)
    {
      U8 index;     /* check whether a task is blocked */
              /* on this mailbox */
      NUSE_Mailbox_Blocking_Count[mailbox]--;
      for (index = 0; index < NUSE_TASK_NUMBER; index++)
      {
        if ((LONIB(NUSE_Task_Status[index]) ==
                                   NUSE_MAILBOX_SUSPEND) &&
          (HINIB(NUSE_Task_Status[index]) == mailbox))
        {
          NUSE_Task_Blocking_Return[index] = NUSE_SUCCESS;
          NUSE_Wake_Task(index);
          break;
        }
      }
    }
    return_value = NUSE_SUCCESS;
    suspend = NUSE_NO_SUSPEND;
  }
  else /* mailbox empty */
  {
    if (suspend == NUSE_NO_SUSPEND)
    {
      return_value = NUSE_MAILBOX_EMPTY;
    }
    else
    {                        /* block task */
      NUSE_Mailbox_Blocking_Count[mailbox]++;
      NUSE_Suspend_Task(NUSE_Task_Active, (mailbox << 4) |
                                   NUSE_MAILBOX_SUSPEND);
      return_value =
      NUSE_Task_Blocking_Return[NUSE_Task_Active];
      if (return_value! = NUSE_SUCCESS)
      {
        suspend = NUSE_NO_SUSPEND;
      }
    }
  }
} while (suspend == NUSE_SUSPEND);
```

Some explanation may be useful:

The code is enclosed in a do...while loop, which continues while the parameter suspend has the value NUSE_SUSPEND.

If the mailbox is full, the stored message is returned, and the mailbox status changed to indicate that it is empty. A check is made on whether any tasks are suspended (waiting to send) on the mailbox. If there are any tasks waiting, the first one is woken. The suspend variable is set to NUSE_NO_SUSPEND and the API call exits with NUSE_SUCCESS.

If the mailbox is empty and suspend is set to NUSE_NO_SUSPEND, the API call exits with NUSE_MAILBOX_EMPTY. If suspend was set to NUSE_SUSPEND, the task is suspended. On return (i.e., when the task is woken up), if the return value is NUSE_SUCCESS, indicating that the task was woken because a message had been sent (as opposed to a reset of the mailbox) the code loops back to the top.

Mailbox utility services

Nucleus RTOS has four API calls that provide utility functions associated with mailboxes: reset mailbox, return information about a mailbox, return number of mailboxes in the application, and return pointers to all mailboxes in the application. The first three of these are implemented in Nucleus SE.

Resetting a mailbox

This API call restores the mailbox to its initial, unused state. Any message stored in the mailbox is lost. Any tasks that were suspended on the mailbox are resumed and receive a return code of NUSE_MAILBOX_WAS_RESET.

Nucleus RTOS API call for resetting a mailbox

Service call prototype:

```
STATUS NU_Reset_Mailbox(NU_MAILBOX *mailbox);
```

Parameters:

mailbox—pointer to the user-supplied mailbox control block

Returns:

NU_SUCCESS—The call was completed successfully.

NU_INVALID_MAILBOX—The mailbox pointer is not valid.

Nucleus SE API call for resetting a mailbox

This API call supports the key functionality of the Nucleus
RTOS API.

Service call prototype:

```
STATUS NUSE_Mailbox_Reset(NUSE_MAILBOX mailbox);
```

Parameters:
mailbox—the index (ID) of the mailbox to be reset

Returns:

NUSE_SUCCESS—The call was completed successfully.

NUSE_INVALID_MAILBOX—The mailbox index is not valid.

Nucleus SE implementation of mailbox reset

The bulk of the code of the NUSE_Mailbox_Reset() API function—
after parameter checking—is selected by conditional compilation,
dependent on whether support for blocking (task suspend) API calls
is enabled. We will look at the two variants separately here.

If blocking is not enabled, the API function code is almost
trivial. The mailbox is marked as unused by setting its entry in
NUSE_Mailbox_Status[] to FALSE.

When blocking is enabled, the code becomes more complex:

```
while (NUSE_Mailbox_Blocking_Count[mailbox] != 0)
{
  U8 index;    /* check whether any tasks are blocked */
          /* on this mailbox */

  for (index = 0; index < NUSE_TASK_NUMBER; index++)
  {
    if ((LONIB(NUSE_Task_Status[index]) ==
                          NUSE_MAILBOX_SUSPEND)
      && (HINIB(NUSE_Task_Status[index]) == mailbox))
    {
      NUSE_Task_Blocking_Return[index] =
                          NUSE_MAILBOX_WAS_RESET;
      NUSE_Task_Status[index] = NUSE_READY;
      break;
    }
  }
  NUSE_Mailbox_Blocking_Count[mailbox]--;
}

#if NUSE_SCHEDULER_TYPE == NUSE_PRIORITY_SCHEDULER
  NUSE_Reschedule(NUSE_NO_TASK);
#endif
```

Initially the mailbox is marked as empty.

Each task suspended on the mailbox is marked as *Ready* with a suspend return code of `NUSE_MAILBOX_WAS_RESET`. After this process is complete, if the Priority scheduler is in use, a call is made to `NUSE_Reschedule()`, as one or more higher priority tasks may have been readied and needs to be allowed to run.

Mailbox information

This service call obtains a selection of information about a mailbox. The Nucleus SE implementation differs from Nucleus RTOS in that it returns less information, as object naming and suspend ordering are not supported and task suspend may not be enabled.

Nucleus RTOS API call for mailbox information

Service call prototype:

```
STATUS NU_Mailbox_Information(
NU_MAILBOX *mailbox,
CHAR *name,
OPTION *suspend_type,
DATA_ELEMENT *message_present,
UNSIGNED *tasks_waiting,
NU_TASK **first_task);
```

Parameters:

mailbox—pointer to the user-supplied mailbox control block;

name—pointer to an 8-character destination area for the mailbox's name. This includes space for a NULL terminator;

suspend_type—pointer to a variable for holding the task suspend type. Valid task suspend types are NU_FIFO and NU_PRIORITY;

message_present—a pointer to a variable, which will receive a value of NU_TRUE or NU_FALSE depending on whether the mailbox is full or not;

tasks_waiting—a pointer to a variable, which will receive the number of tasks suspended on this mailbox;

first_task—a pointer to a task pointer, which will receive the pointer to the first suspended task.

Returns:

NU_SUCCESS—The call was completed successfully.

NUSE_INVALID_MAILBOX—The mailbox pointer is not valid.

Nucleus SE API call for mailbox information

This API call supports the key functionality of the Nucleus RTOS API.

Service call prototype:

```
STATUS NUSE_Mailbox_Information(
NUSE_MAILBOX mailbox,
U8 *message_present,
U8 *tasks_waiting,
NUSE_TASK *first_task);
```

Parameters:

mailbox—the index of the mailbox about which information is being requested;

message_present—a pointer to a variable, which will receive a value of TRUE or FALSE depending on whether the mailbox is full or not;

tasks_waiting—a pointer to a variable, which will receive the number of tasks suspended on this mailbox (nothing returned if task suspend is disabled);

first_task—a pointer to a variable of type NUSE_TASK, which will receive the index of the first suspended task (nothing returned if task suspend is disabled).

Returns:

NUSE_SUCCESS—The call was completed successfully.

NUSE_INVALID_MAILBOX—The mailbox index is not valid.

NUSE_INVALID_POINTER—One or more of the pointer parameters is invalid.

Nucleus SE implementation of mailbox information

The implementation of this API call is quite straightforward:

```
*message_present = NUSE_Mailbox_Status[mailbox];

#if NUSE_BLOCKING_ENABLE

  *tasks_waiting = NUSE_Mailbox_Blocking_Count[mailbox];
  if (NUSE_Mailbox_Blocking_Count[mailbox]!= 0)
  {
    U8 index;

    for (index = 0; index < NUSE_TASK_NUMBER; index++)
```

```
    {
      if ((LONIB(NUSE_Task_Status[index]) ==
                              NUSE_MAILBOX_SUSPEND)
        && (HINIB(NUSE_Task_Status[index]) == mailbox))
      {
        *first_task = index;
        break;
      }
    }
  }
  else
  {
    *first_task = 0;
  }

#else

  *tasks_waiting = 0;
  *first_task = 0;

#endif

return NUSE_SUCCESS;
```

The function returns the mailbox status. Then, if blocking API calls is enabled, the number of waiting tasks and the index of the first one are returned (otherwise, these two parameters are set to 0).

Obtaining the number of mailboxes

This service call returns the number of mailboxes configured in the application. Whilst in Nucleus RTOS this will vary over time and the returned value will represent the current number of mailboxes, in Nucleus SE the value returned is set at build time and cannot change.

Nucleus RTOS API call for mailbox count

Service call prototype:

```
UNSIGNED NU_Established_Mailboxes(VOID);
```

Parameters:

> None

Returns:

> The number of created mailboxes in the application

Nucleus SE API call for mailbox count

This API call supports the key functionality of the Nucleus RTOS API.

Service call prototype:

```
U8 NUSE_Mailbox_Count(void);
```

Parameters:

None

Returns:

The number of configured mailboxes in the application

Nucleus SE implementation of mailbox count

The implementation of this API call is almost trivially simple: the value of the `#define` symbol `NUSE_MAILBOX_NUMBER` is returned.

Data structures

Mailboxes utilize two or three data structures—all in RAM—which, like other Nucleus SE objects, are a series of tables, included and dimensioned according to the number of mailboxes configured and options selected.

I strongly recommend that application code does not access these data structures directly but uses the provided API functions. This avoids incompatibility with future versions of Nucleus SE and unwanted side effects and simplifies porting of an application to Nucleus RTOS. The details of data structures are included here to facilitate easier understanding of the working of the service call code and for debugging.

RAM data

These data structures are:

`NUSE_Mailbox_Data[]`—This is an array of type `ADDR`, with one entry for each configured mailbox and is where the mailbox data is stored.

`NUSE_Mailbox_Status[]`—This is an array of type `U8`, with one entry for each configured mailbox, which tracks the usage of the mailboxes. A nonzero value (`TRUE`) indicates that a mailbox is full.

`NUSE_Mailbox_Blocking_Count[]`—This type `U8` array contains the counts of how many tasks are blocked on each mailbox. This array only exists if blocking API call support is enabled.

These data structures are all initialized to zeros by `NUSE_Init_Mailbox()` when Nucleus SE starts up. This is logical, as it renders every mailbox as being empty (unused). Chapter 17, *Nucleus SE Initialization and Start-up*, provides a full description of Nucleus SE start-up procedures.

Here are the definitions of these data structures in `nuse_init.c` file.

```
RAM ADDR NUSE_Mailbox_Data[NUSE_MAILBOX_NUMBER];
RAM U8 NUSE_Mailbox_Status[NUSE_MAILBOX_NUMBER];

#if NUSE_BLOCKING_ENABLE

  RAM U8 NUSE_Mailbox_Blocking_Count[NUSE_MAILBOX_NUMBER];

#endif
```

ROM data

There are no ROM data structures associated with mailboxes.

Mailbox data footprint

Like all kernel objects in Nucleus SE, the amount of data memory required for mailboxes is readily predictable.

The ROM data footprint for all the mailboxes in an application is 0.

The RAM data footprint (in bytes) for all the mailboxes in an application, when blocking API calls is enabled, may be computed thus:

```
NUSE_MAILBOX_NUMBER * (sizeof(ADDR) + 2)
```

Otherwise, it is:

```
NUSE_MAILBOX_NUMBER * (sizeof(ADDR) + 1)
```

Unimplemented API calls

Four mailbox API calls found in Nucleus RTOS are not implemented in Nucleus SE:

Create mailbox

This API call creates a mailbox. It is not needed with Nucleus SE, as mailboxes are created statically.

Service call prototype:

```
STATUS NU_Create_Mailbox(
NU_MAILBOX *mailbox,
CHAR *name,
OPTION suspend_type);
```

Parameters:

`mailbox`—Pointer to a user-supplied mailbox control block; this will be used as a "handle" for the mailbox in other API calls.

`name`—Pointers to a 7-character, `NULL`-terminated name for the mailbox.

`suspend_type`—Specifies how tasks suspend on the mailbox. Valid options for this parameter are `NU_FIFO` and `NU_PRIORITY`, which represent First-In-First-Out (FIFO) and priority-order task suspension, respectively.

Returns:

`NU_SUCCESS`—indicates successful completion of the service.

`NU_INVALID_MAILBOX`—indicates the mailbox control block pointer is `NULL` or already in use.

`NU_INVALID_SUSPEND`—indicates that the `suspend_type` parameter is invalid.

Delete mailbox

This API call deletes a previously created mailbox. It is not needed with Nucleus SE, as mailboxes are created statically and cannot be deleted.

Service call prototype:

```
STATUS NU_Delete_Mailbox(NU_MAILBOX *mailbox);
```

Parameters:

`mailbox`—pointer to mailbox control block

Returns:

`NU_SUCCESS`—indicates successful completion of the service.
`NU_INVALID_MAILBOX`—indicates the mailbox pointer is invalid.

Mailbox pointers

This API call builds a sequential list of pointers to all mailboxes in the system. It is not needed with Nucleus SE, as mailboxes are identified by a simple index, not a pointer, and it would be redundant.

Service call prototype:

```
UNSIGNED NU_Mailbox_Pointers(
NU_MAILBOX **pointer_list,
UNSIGNED maximum_pointers);
```

Parameters:

pointer_list—pointer to an array of NU_MAILBOX pointers; this array will be filled with pointers to established mailboxes in the system;

maximum_pointers—the maximum number of pointers to place in the array.

Returns:

The number of NU_MAILBOX pointers placed into the array

Broadcast to mailbox

This API call broadcasts a message to all tasks waiting for a message from the specified mailbox. It is not implemented with Nucleus SE, as it would have added excessive complexity.

Service call prototype:

```
STATUS NU_Broadcast_To_Mailbox(
NU_MAILBOX *mailbox,
VOID *message,
UNSIGNED suspend);
```

Parameters:

mailbox—pointer to mailbox control block;

message—pointer to the broadcast message;

suspend—specifies whether or not to suspend the calling task if the mailbox already contains a message; valid options for this parameter are NU_NO_SUSPEND, NU_SUSPEND, or a timeout value.

Returns:

NU_SUCCESS—indicates successful completion of the service.

NU_INVALID_MAILBOX—indicates the mailbox pointer is invalid.

NU_INVALID_POINTER—indicates that the message pointer is NULL.

NU_INVALID_SUSPEND—indicates that suspend attempted from a nontask thread.

NU_MAILBOX_FULL—indicates the mailbox already contains a message.

NU_TIMEOUT—indicates the mailbox is still full after the time-out has expired.

NU_MAILBOX_DELETED—mailbox was deleted, while the task was suspended.

NU_MAILBOX_RESET—mailbox was reset, while the task was suspended.

Compatibility with Nucleus RTOS

With all aspects of Nucleus SE, it was my goal to maintain as high a level of applications code compatibility with Nucleus RTOS as possible. Mailboxes are no exception, and, from a user's perspective, they are implemented in much the same way as in Nucleus RTOS. There are areas of incompatibility, which have come about where I determined that such an incompatibility would be acceptable, given that the resulting code is easier to understand, or, more likely, could be made more memory efficient. Otherwise, Nucleus RTOS API calls may be almost directly mapped onto Nucleus SE calls. Chapter 20, *Using Nucleus SE*, includes further information on using Nucleus SE for users of Nucleus RTOS.

Object identifiers

In Nucleus RTOS, all objects are described by a data structure—a control block—which has a specific data type. A pointer to this control block serves as an identifier for the mailbox. In Nucleus SE, I decided that a different approach was needed for memory efficiency, and all kernel objects are described by a number of tables in RAM and/or ROM. The size of these tables is determined by the number of each object type that is configured. The identifier for a specific object is simply an index into those tables. So, I have defined NUSE_MAILBOX as being equivalent to U8; a variable—not a pointer—of this type then serves as the mailbox identifier. This is a small incompatibility, which is easily handled if code is ported to or from Nucleus RTOS. Object identifiers are normally just stored and passed around and not operated on in any way.

Nucleus RTOS also supports naming of mailboxes. These names are only used for target-based debug facilities. I omitted them from Nucleus SE to save memory.

Message size and type

In Nucleus RTOS, a mailbox message consists of four 32-bit words. I decided to reduce this to a single variable of type ADDR in Nucleus SE. This change imparts significant increases in memory and execution time efficiency. It also recognizes that a common application for a mailbox is to send a pointer to something from one task to another. This incompatibility will present few challenges when porting application code to Nucleus RTOS. Nucleus SE could be modified quite readily, if a different message format were required.

Unimplemented API calls

Nucleus RTOS supports nine service calls to work with mailboxes. Of these, four are not implemented in Nucleus SE. Details of these and of the decision to omit them may be found in "*Unimplemented API calls*" section earlier in this chapter.

12

Queues

Queues were introduced in Chapter 3, *RTOS Services and Facilities*. They provide a more flexible means of passing simple messages between tasks than mailboxes.

Using queues

In Nucleus SE, queues are configured at build time. There may be a maximum of 16 queues configured for an application. If no queues are configured, no data structures or service call code appertaining to queues are included in the application.

A queue is simply a set of storage locations, each big enough to hold a single data item of type **ADDR**, access to which is controlled so that it may be safely utilized by multiple tasks. Tasks can write to a queue repeatedly until all the locations are full. Tasks can read from a queue and data is normally received on a first-in, first-out (FIFO) basis. Trying to send to a full queue or read from an empty one may result in an error or task suspension, depending on options selected in the application program interface (API) call and the Nucleus SE configuration.

Queues and pipes

Nucleus SE also supports pipes, which were also introduced in Chapter 3, *RTOS Services and Facilities*, and are covered in detail in Chapter 13, *Pipes*. The main difference between queues and pipes is the message size. Queues carry messages comprise a single **ADDR**—these would commonly be pointers. A pipe carries messages that are an arbitrary number of bytes long; the size is fixed for each pipe in the application and set at configuration time.

Configuring queues

Number of queues

As with most aspects of Nucleus SE, the configuration of queues is primarily controlled by `#define` statements in `nuse_config.h`. The

Embedded RTOS Design. DOI: https://doi.org/10.1016/B978-0-12-822851-7.00012-6
© 2021 Elsevier Ltd. All rights reserved.

key setting is NUSE_QUEUE_NUMBER, which determines how many queues are configured for the application. The default setting is 0 (i.e., no queues are in use) and you can set it to any value up to 16. An erroneous value will result in a compile time error, which is generated by a test in nuse_config_check.h (this is included into nuse_config.c and hence compiled with this module), resulting in a #error statement being compiled.

Choosing a nonzero value is the "master enable" for queues. This results in some data structures being defined and sized accordingly, of which more later. It also activates the API enabling settings.

API enables

Every API function (service call) in Nucleus SE has an enabling #define symbol in nuse_config.h. For queues, these are:

```
NUSE_QUEUE_SEND
NUSE_QUEUE_RECEIVE
NUSE_QUEUE_JAM
NUSE_QUEUE_RESET
NUSE_QUEUE_INFORMATION
NUSE_QUEUE_COUNT
```

By default, all of these are set to FALSE, thus disabling each service call and inhibiting the inclusion of any implementation code. To configure queues for an application, you need to select the API calls that you want to use and set their enabling symbols to TRUE.

Here is an extract from the default nuse_config.h file.

```
#define NUSE_QUEUE_NUMBER   0   /* Number of queues in the
                 system - 0-16 */

                /* Service call enablers */
#define NUSE_QUEUE_SEND     FALSE
#define NUSE_QUEUE_RECEIVE      FALSE
#define NUSE_QUEUE_JAM      FALSE
#define NUSE_QUEUE_RESET      FALSE
#define NUSE_QUEUE_INFORMATION      FALSE
#define NUSE_QUEUE_COUNT      FALSE
```

A compile time error will result if a queue API function is enabled and no queues are configured (except for NUSE_Queue_Count() which is always permitted). If your code uses an API call, which has not been enabled, a link time error will result, as no implementation code will have been included in the application.

Queue service calls

Nucleus RTOS supports 10 service calls that appertain to queues, which provide the following functionality:

Send a message to a queue. Implemented by `NUSE_Queue_Send()` in Nucleus SE.

Receive a message from a queue. Implemented by `NUSE_Queue_Receive()` in Nucleus SE.

Send a message to the front of a queue. Implemented by `NUSE_Queue_Jam()` in Nucleus SE.

Restore a queue to the unused state, with no tasks suspended (reset). Implemented by `NUSE_Queue_Reset()` in Nucleus SE.

Provide information about a specified queue. Implemented by `NUSE_Queue_Information()` in Nucleus SE.

Return a count of how many queues are (currently) configured for the application. Implemented by `NUSE_Queue_Count()` in Nucleus SE.

Add a new queue to the application (create). Not implemented in Nucleus SE.

Remove a queue from the application (delete). Not implemented in Nucleus SE.

Return pointers to all the queues (currently) in the application. Not implemented in Nucleus SE.

Send a message to all the tasks that are suspended on a queue (broadcast). Not implemented in Nucleus SE.

The implementation of each of these service calls will be examined in detail.

Queue write and read services

The fundamental operations, which can be performed on a queue, are writing data to it—which is sometimes termed *sending*—and reading data from it—which is also termed *receiving*. It is also possible to write data to the front of a queue—which is also termed *jamming*. Nucleus RTOS and Nucleus SE each provide three basic API calls for these operations, which will be discussed here.

Writing to a queue

The Nucleus RTOS API call for writing to a queue is very flexible, enabling you to suspend indefinitely, or with a timeout, if the operation cannot be completed immediately; that is, you try to write to a full queue. Nucleus SE provides the same service, except task suspend is optional and timeout is not implemented.

Nucleus RTOS also offers a facility to broadcast to a queue, but this is not supported by Nucleus SE. It is described under "Unimplemented API calls" later in this chapter.

Nucleus RTOS API call for sending to a queue

Service call prototype:

```
STATUS NU_Send_To_Queue(
NU_QUEUE *queue,
VOID *message,
UNSIGNED size,
UNSIGNED suspend);
```

Parameters:

queue—pointer to the user-supplied queue control block;

message—a pointer to the message to be sent;

size—the number of UNSIGNED data elements in the message. If the queue supports variable-length messages, this parameter must be equal to or less than the message size supported by the queue. If the queue supports fixed-size messages, this parameter must be exactly the same as the message size supported by the queue;

suspend—specification for task suspend; may be NU_NO_SUSPEND or NU_SUSPEND or a timeout value.

Returns:

NU_SUCCESS—The call was completed successfully.

NU_INVALID_QUEUE—The queue pointer is invalid.

NU_INVALID_POINTER—The message pointer is NULL.

NU_INVALID_SIZE—The message size is incompatible with the message size supported by the queue.

NU_INVALID_SUSPEND—Suspend was attempted from a non task thread.

NU_QUEUE_FULL—The queue is full and suspend was not specified.

NU_TIMEOUT—the queue is still full even after suspending for the specified timeout value.

NU_QUEUE_DELETED—The queue was deleted, while the task was suspended.

NU_QUEUE_RESET—The queue was reset, while the task was suspended.

Nucleus SE API call for sending to a queue

This API call supports the key functionality of the Nucleus RTOS API.

Service call prototype:

```
STATUS NUSE_Queue_Send(
NUSE_QUEUE queue,
ADDR *message,
U8 suspend);
```

Parameters:

queue—the index (ID) of the queue to be utilized;

message—a pointer to the message to be sent, which is a single variable of type ADDR;

suspend—specification for task suspend; may be NUSE_NO_SUSPEND or NUSE_SUSPEND.

Returns:

NUSE_SUCCESS—The call was completed successfully.

NUSE_INVALID_QUEUE—The queue index is invalid.

NUSE_INVALID_POINTER—The message pointer is NULL.

NUSE_INVALID_SUSPEND—Suspend was attempted from a non-task thread or when blocking API calls were not enabled.

NUSE_QUEUE_FULL—The queue is full and suspend was not specified.

NUSE_QUEUE_WAS_RESET—The queue was reset, while the task was suspended.

Nucleus SE implementation of queue send

The bulk of the code of the `NUSE_Queue_Send()` API function—after parameter checking—is selected by conditional compilation, dependent on whether support for blocking (task suspend) API calls is enabled. We will look at the two variants separately here.

If blocking is not enabled, the code for this API call is quite simple:

```
if (NUSE_Queue_Items[queue] == NUSE_Queue_Size[queue])
                                            /* queue full */
{
  return_value = NUSE_QUEUE_FULL;
}
else              /* queue element available */
{
  NUSE_Queue_Data[queue][NUSE_Queue_Head[queue]++] =
                                            *message;
  if (NUSE_Queue_Head[queue] == NUSE_Queue_Size[queue])
  {
    NUSE_Queue_Head[queue] = 0;
  }
  NUSE_Queue_Items[queue]++;
  return_value = NUSE_SUCCESS;
}
```

The function simply checks that there is room in the queue and uses the `NUSE_Queue_Head[]` index to store the message in the queue's data area.

When blocking is enabled, the code becomes more complex:

```
do
{
  if (NUSE_Queue_Items[queue] == NUSE_Queue_Size[queue])
                                            /* queue full */
  {
    if (suspend == NUSE_NO_SUSPEND)
    {
      return_value = NUSE_QUEUE_FULL;
    }
    else
    {              /* block task */
      NUSE_Queue_Blocking_Count[queue]++;
      NUSE_Suspend_Task(NUSE_Task_Active,(queue << 4) |
                                NUSE_QUEUE_SUSPEND);
      return_value =
      NUSE_Task_Blocking_Return[NUSE_Task_Active];
      if (return_value! = NUSE_SUCCESS)
```

```
        {
          suspend = NUSE_NO_SUSPEND;
        }
      }
    }
    else
    {                    /* queue element available */
      NUSE_Queue_Data[queue][NUSE_Queue_Head[queue]++] =
                                              *message;
      if (NUSE_Queue_Head[queue] == NUSE_Queue_Size[queue])
      {
        NUSE_Queue_Head[queue] = 0;
      }
      NUSE_Queue_Items[queue]++;
      if (NUSE_Queue_Blocking_Count[queue]!= 0)
      {
        U8 index;        /* check whether a task is blocked on this
                                                      queue */

        NUSE_Queue_Blocking_Count[queue]--;
        for (index = 0; index < NUSE_TASK_NUMBER; index++)
        {
          if ((LONIB(NUSE_Task_Status[index]) ==
          NUSE_QUEUE_SUSPEND)
            && (HINIB(NUSE_Task_Status[index]) == queue))
          {
            NUSE_Task_Blocking_Return[index] = NUSE_SUCCESS;
            NUSE_Wake_Task(index);
            break;
          }
        }
      }
      return_value = NUSE_SUCCESS;
      suspend = NUSE_NO_SUSPEND;
    }
  } while (suspend == NUSE_SUSPEND);
```

Some explanation of the code may be useful:

The code is enclosed in a `do...while` loop, which continues while the parameter `suspend` has the value `NUSE_SUSPEND`.

If the queue is full and `suspend` is set to `NUSE_NO_SUSPEND`, the API call exits with `NUSE_QUEUE_FULL`. If `suspend` was set to `NUSE_SUSPEND`, the task is suspended. On return (i.e., when the task is woken up), if the return value is `NUSE_SUCCESS`, indicating that the task was woken because a message had been read (as opposed to a reset of the queue), the code loops back to the top.

If the queue is not full, the supplied message is stored using the `NUSE_Queue_Head[]` index to store the message in the queue's data area. A check is made on whether any tasks are suspended (waiting to receive) on the queue. If there are any tasks waiting, the first one is woken. The **suspend** variable is set to **NUSE_NO_SUSPEND** and the API call exits with **NUSE_SUCCESS**.

Reading from a queue

The Nucleus RTOS API call for reading from a queue is very flexible, enabling you to suspend indefinitely, or with a timeout, if the operation cannot be completed immediately; that is, you try to read from an empty queue. Nucleus SE provides the same service, except task suspend is optional and timeout is not implemented.

Nucleus RTOS API call for receiving from a queue

Service call prototype:

```
STATUS NU_Receive_From_Queue(
NU_QUEUE *queue,
VOID *message,
UNSIGNED size,
UNSIGNED *actual_size,
UNSIGNED suspend);
```

Parameters:

queue—pointer to user-supplied queue control block;

message—pointer to storage for the message to be received;

size—the number of **UNSIGNED** data elements in the message. This number must correspond to the message size defined when the queue was created;

suspend—specification for task suspend; may be **NU_NO_SUSPEND** or **NU_SUSPEND** or a timeout value.

Returns:

NU_SUCCESS—The call was completed successfully.

NU_INVALID_QUEUE—The queue pointer is invalid.

NU_INVALID_POINTER—The message pointer is **NULL**.

NU_INVALID_SUSPEND—Suspend was attempted from a nontask thread.

NU_QUEUE_EMPTY—The queue is empty and suspend was not specified.

NU_TIMEOUT—indicates that the queue is still empty even after suspending for the specified timeout value.

`NU_QUEUE_DELETED`—The queue was deleted, while the task was suspended.

`NU_QUEUE_RESET`—The queue was reset, while the task was suspended.

Nucleus SE API call for receiving from a queue

This API call supports the key functionality of the Nucleus RTOS API.

Service call prototype:

```
STATUS NUSE_Queue_Receive(
NUSE_QUEUE queue,
ADDR *message,
U8 suspend);
```

Parameters:

`queue`—the index (ID) of the queue to be utilized;

`message`—a pointer to storage for the message to be received, which is a single variable of type `ADDR`;

`suspend`—specification for task suspend; may be `NUSE_NO_SUSPEND` or `NUSE_SUSPEND`.

Returns:

`NUSE_SUCCESS`—The call was completed successfully.

`NUSE_INVALID_QUEUE`—The queue index is invalid.

`NUSE_INVALID_POINTER`—The message pointer is `NULL`.

`NUSE_INVALID_SUSPEND`—Suspend was attempted from a non-task thread or when blocking API calls were not enabled.

`NUSE_QUEUE_EMPTY`—The queue is empty and suspend was not specified.

`NUSE_QUEUE_WAS_RESET`—The queue was reset, while the task was suspended.

Nucleus SE implementation of queue receive

The bulk of the code of the `NUSE_Queue_Receive()` API function—after parameter checking—is selected by conditional compilation, dependent on whether support for blocking (task suspend) API calls is enabled. We will look at the two variants separately here.

If blocking is not enabled, the code for this API call is quite simple:

```
if (NUSE_Queue_Items[queue] == 0)   /* queue empty */
{
  return_value = NUSE_QUEUE_EMPTY;
}
else
{               /* message available */
  *message =
  NUSE_Queue_Data[queue][NUSE_Queue_Tail[queue]++];
  if (NUSE_Queue_Tail[queue] == NUSE_Queue_Size[queue])
  {
    NUSE_Queue_Tail[queue] = 0;
  }
  NUSE_Queue_Items[queue]--;
  return_value = NUSE_SUCCESS;
}
```

The function simply checks that there is a message in the queue and uses the NUSE_Queue_Tail[] index to obtain the message from the queue's data area and returns the data via the message pointer.

When blocking is enabled, the code becomes more complex:

```
do
{
  if (NUSE_Queue_Items[queue] == 0)   /* queue empty */
  {
   if (suspend == NUSE_NO_SUSPEND)
   {
     return_value = NUSE_QUEUE_EMPTY;
   }
   else
   {                /* block task */
     NUSE_Queue_Blocking_Count[queue]++;
     NUSE_Suspend_Task(NUSE_Task_Active, (queue << 4) |
                                NUSE_QUEUE_SUSPEND);
     return_value =
     NUSE_Task_Blocking_Return[NUSE_Task_Active];
     if (return_value! = NUSE_SUCCESS)
     {
        suspend = NUSE_NO_SUSPEND;
     }
   }
  }
  else
```

```
   {                    /* message available */
   *message =
   NUSE_Queue_Data[queue][NUSE_Queue_Tail[queue]++];
   if (NUSE_Queue_Tail[queue] == NUSE_Queue_Size[queue])
   {
     NUSE_Queue_Tail[queue] = 0;
   }
   NUSE_Queue_Items[queue]--;
   if (NUSE_Queue_Blocking_Count[queue]!= 0)
   {
     U8 index;          /* check whether a task is blocked */
                        /* on this queue */
     NUSE_Queue_Blocking_Count[queue]--;
     for (index = 0; index < NUSE_TASK_NUMBER; index++)
     {
       if ((LONIB(NUSE_Task_Status[index]) ==
                         NUSE_QUEUE_SUSPEND)
         && (HINIB(NUSE_Task_Status[index]) == queue))
       {
         NUSE_Task_Blocking_Return[index] = NUSE_SUCCESS;
         NUSE_Wake_Task(index);
         break;
       }
     }
   }
   return_value = NUSE_SUCCESS;
   suspend = NUSE_NO_SUSPEND;
  }
} while (suspend == NUSE_SUSPEND);
```

Some explanation of the code may be useful:

The code is enclosed in a `do...while` loop, which continues while the parameter `suspend` has the value `NUSE_SUSPEND`.

If the queue is empty and `suspend` is set to `NUSE_NO_SUSPEND`, the API call exits with `NUSE_QUEUE_EMPTY`. If `suspend` was set to `NUSE_SUSPEND`, the task is suspended. On return (i.e., when the task is woken up), if the return value is `NUSE_SUCCESS`, indicating that the task was woken because a message had been sent (as opposed to a reset of the queue), the code loops back to the top.

If the queue contains any messages, a stored message is returned using the `NUSE_Queue_Tail[]` index to obtain the message from the queue's data area. A check is made on whether any tasks are suspended (waiting to send) on the queue. If there are any tasks waiting, the first one is woken. The `suspend` variable is set to `NUSE_NO_SUSPEND` and the API call exits with `NUSE_SUCCESS`.

Writing to the front of a queue

The Nucleus RTOS API call for writing to the front of a queue is very flexible, enabling you to suspend indefinitely, or with a timeout, if the operation cannot be completed immediately; that is, you try to write to a full queue. Nucleus SE provides the same service, except task suspend is optional and timeout is not implemented.

Nucleus RTOS API call for jamming to a queue

Service call prototype:

```
STATUS NU_Send_To_Front_Of_Queue(
NU_QUEUE *queue,
VOID *message,
UNSIGNED size,
UNSIGNED suspend);
```

Parameters:

queue—pointer to a user-supplied queue control block;

message—a pointer to the message to be sent;

size—the number of UNSIGNED data elements in the message. If the queue supports variable-length messages, this parame-ter must be equal to or less than the message size supported by the queue. If the queue supports fixed-size messages, this parameter must be exactly the same as the message size supported by the queue;

suspend—specification for task suspend; may be NU_NO_SUSPEND or NU_SUSPEND or a timeout value.

Returns:

NU_SUCCESS—The call was completed successfully.

NU_INVALID_QUEUE—The queue pointer is invalid.

NU_INVALID_POINTER—The message pointer is NULL.

NU_INVALID_SIZE—The message size is incompatible with the message size supported by the queue.

NU_INVALID_SUSPEND—Suspend was attempted from a non-task thread.

NU_QUEUE_FULL—The queue is full and suspend was not specified.

NU_TIMEOUT—The queue is still full even after suspending for the specified timeout value.

NU_QUEUE_DELETED—The queue was deleted, while the task was suspended.

NU_QUEUE_RESET—The queue was reset, while the task was suspended.

Nucleus SE API call for jamming to a queue

This API call supports the key functionality of the Nucleus RTOS API.

Service call prototype:

```
STATUS NUSE_Queue_Jam(
NUSE_QUEUE queue,
ADDR *message,
U8 suspend);
```

Parameters:

queue—the index (ID) of the queue to be utilized;

message—a pointer to the message to be sent, which is a single variable of type ADDR;

suspend—specification for task suspend; may be NUSE_NO_SUSPEND or NUSE_SUSPEND.

Returns:

NUSE_SUCCESS—The call was completed successfully.

NUSE_INVALID_QUEUE—The queue index is invalid.

NUSE_INVALID_POINTER—The message pointer is NULL.

NUSE_INVALID_SUSPEND—Suspend was attempted from a nontask thread or when blocking API calls were not enabled.

NUSE_QUEUE_FULL—The queue is full and suspend was not specified.

NUSE_QUEUE_WAS_RESET—The queue was reset, while the task was suspended.

Nucleus SE implementation of jamming to a queue

The bulk of the code of the NUSE_Queue_Jam() API function is very similar to that of NUSE_Queue_Send(), except that the data are stored using the NUSE_Queue_Tail[] index; thus:

```
if (NUSE_Queue_Items[queue] == NUSE_Queue_Size[queue]) /*
                                            queue full */
{
  return_value = NUSE_QUEUE_FULL;
}
else        /* queue element available */
{
  if (NUSE_Queue_Tail[queue] == 0)
  {
    NUSE_Queue_Tail[queue] = NUSE_Queue_Size[queue] - 1;
  }
  else
  {
    NUSE_Queue_Tail[queue]--;
  }
  NUSE_Queue_Data[queue][NUSE_Queue_Tail[queue]] = *message;
  NUSE_Queue_Items[queue]++;
  return_value = NUSE_SUCCESS;
}
```

Queue utility services

Nucleus RTOS has four API calls that provide utility functions associated with queues: reset queue, return information about a queue, return number of queues in the application, and return pointers to all queues in the application. The first three of these are implemented in Nucleus SE.

Resetting a queue

This API call restores the queue to its initial, unused state. Any messages stored in the queue are lost. Any tasks that were suspended on the queue are resumed and receive a return code of NUSE_QUEUE_WAS_RESET.

Nucleus RTOS API call for resetting a queue

Service call prototype:

STATUS NU_Reset_Queue(NU_QUEUE *queue);

Parameters:

queue—pointer to user-defined queue control block

Returns:

NU_SUCCESS—The call was completed successfully.

NU_INVALID_QUEUE—The queue pointer is not valid.

Nucleus SE API call for resetting a queue

This API call supports the key functionality of the Nucleus RTOS API.

Service call prototype:

```
STATUS NUSE_Queue_Reset(NUSE_QUEUE queue);
```

Parameters:

queue—the index (ID) of the queue to be reset

Returns:

NUSE_SUCCESS—The call was completed successfully.

NUSE_INVALID_QUEUE—The queue index is not valid.

Nucleus SE implementation of queue reset

The initial part of the code of the NUSE_Queue_Reset() API function—after parameter checking—is quite straightforward. The head and tail indexes and the queue's message count are all set to zero.

When blocking is enabled, additional code takes care of waking up any suspended tasks; thus:

```
while (NUSE_Queue_Blocking_Count[queue] != 0)
{
  U8 index;              /* check whether any tasks are blocked */
                /* on this queue */

  for (index = 0; index < NUSE_TASK_NUMBER; index++)
  {
    if ((LONIB(NUSE_Task_Status[index]) == NUSE_QUEUE_SUSPEND)
      && (HINIB(NUSE_Task_Status[index]) == queue))
    {
      NUSE_Task_Blocking_Return[index] = NUSE_QUEUE_WAS_RESET;
      NUSE_Task_Status[index] = NUSE_READY;
      break;
    }
  }
  NUSE_Queue_Blocking_Count[queue]--;
}

#if NUSE_SCHEDULER_TYPE == NUSE_PRIORITY_SCHEDULER
  NUSE_Reschedule(NUSE_NO_TASK);
#endif
```

Each task suspended on the queue is marked as *Ready* with a suspend return code of NUSE_QUEUE_WAS_RESET. After this process is complete, if the Priority scheduler is in use, a call is made to NUSE_Reschedule(), as one or more higher priority tasks may have been readied and needs to be allowed to run.

Queue information

This service call obtains a selection of information about a queue. The Nucleus SE implementation differs from Nucleus RTOS in that it returns less information, as object naming, variable message size, and suspend ordering are not supported and task suspend may not be enabled.

Nucleus RTOS API call for queue information

Service call prototype:

```
STATUS NU_Queue_Information(
NU_QUEUE *queue,
CHAR *name,
VOID **start_address
UNSIGNED *queue_size,
UNSIGNED *available,
UNSIGNED *messages,
OPTION *message_type,
UNSIGNED *message_size,
OPTION *suspend_type,
UNSIGNED *tasks_waiting,
NU_TASK **first_task);
```

Parameters:

queue—pointer to the user-supplied queue control block;

name—pointer to an 8-character destination area for the message-queue's name;

start_address—a pointer to a pointer, which will receive the address of the start of the queue's data area;

queue_size—a pointer to a variable for holding the total number of UNSIGNED data elements in the queue;

available—a pointer to a variable for holding the num-ber of available UNSIGNED data elements in the queue;

messages—a pointer to a variable for holding the number of messages currently in the queue;

`message_type`—pointer to a variable for holding the type of messages supported by the queue; valid message types are `NU_FIXED_SIZE` and `NU_VARIABLE_SIZE`;

`message_size`—pointer to a variable for holding the num-ber of `UNSIGNED` data elements in each queue message; if the queue supports variable-length messages, this num-ber is the maximum message size;

`suspend_type`—pointer to a variable for holding the task sus-pend type. Valid task suspend types are `NU_FIFO` and `NU_PRIORITY`;

`tasks_waiting`—a pointer to a variable which will receive the number of tasks suspended on this queue;

`first_task`—a pointer to a task pointer; the pointer of the first suspended task is placed in this task pointer.

Returns:

`NU_SUCCESS`—The call was completed successfully.

`NU_INVALID_QUEUE`—The queue pointer is not valid.

Nucleus SE API call for queue information

This API call supports the key functionality of the Nucleus RTOS API.

Service call prototype:

```
STATUS NUSE_Queue_Information(
NUSE_QUEUE queue,
ADDR *start_address,
U8 *queue_size,
U8 *available,
U8 *messages,
U8 *tasks_waiting,
NUSE_TASK *first_task);
```

Parameters:

`queue`—the index of the queue about which information is being requested;

`start_address`—a pointer to a variable of type `ADDR`, which will receive the address of the start of the queue's data area;

`queue_size`—a pointer to a variable of type `U8`, which will receive the total number of messages for which the queue has capacity;

available—a pointer to a variable of type **U8**, which will receive the number of messages for which the queue has currently remaining capacity;

messages—a pointer to a variable of type **U8**, which will receive the number of messages currently in the queue;

tasks_waiting—a pointer to a variable, which will receive the number of tasks suspended on this queue (nothing returned if task suspend is disabled);

first_task—a pointer to a variable of type **NUSE_TASK**, which will receive the index of the first suspended task (nothing returned if task suspend is disabled).

Returns:

NUSE_SUCCESS—The call was completed successfully.

NUSE_INVALID_QUEUE—The queue index is not valid.

NUSE_INVALID_POINTER—One or more of the pointer parameters is invalid.

Nucleus SE implementation of queue information

The implementation of this API call is quite straightforward:

```
*start_address = NUSE_Queue_Data[queue];
*queue_size = NUSE_Queue_Size[queue];
*available = NUSE_Queue_Size[queue] - NUSE_Queue_Items[queue];
*messages = NUSE_Queue_Items[queue];
#if NUSE_BLOCKING_ENABLE

  *tasks_waiting = NUSE_Queue_Blocking_Count[queue];
  if (NUSE_Queue_Blocking_Count[queue] != 0)
  {
    U8 index;

    for (index = 0; index < NUSE_TASK_NUMBER; index++)
    {
      if ((LONIB(NUSE_Task_Status[index]) ==
      NUSE_QUEUE_SUSPEND)
        && (HINIB(NUSE_Task_Status[index]) == queue))
      {
        *first_task = index;
        break;
      }
    }
  }
  else
  {
```

```
    *first_task = 0;
  }

 #else

 *tasks_waiting = 0;
 *first_task = 0;

#endif
```

The function returns the queue status. Then, if blocking API calls is enabled, the number of waiting tasks and the index of the first one are returned (otherwise, these two parameters are set to 0).

Obtaining the number of queues

This service call returns the number of queues configured in the application. Whilst in Nucleus RTOS this will vary over time and the returned value will represent the current number of queues, in Nucleus SE the value returned is set at build time and cannot change.

Nucleus RTOS API call for queue count

Service call prototype:

```
UNSIGNED NU_Established_Queues(VOID);
```

Parameters:

None

Returns:

The number of created queues in the system.

Nucleus SE API call for queue count

This API call supports the key functionality of the Nucleus RTOS API.

Service call prototype:

```
U8 NUSE_Queue_Count(void);
```

Parameters:

None

Returns:

The number of configured queues in the application

Nucleus SE implementation of queue count

The implementation of this API call is almost trivially simple: the value of the `#define` symbol `NUSE_QUEUE_NUMBER` is returned.

Data structures

Queues utilize five or six data structures—all in RAM and ROM—which, like other Nucleus SE objects, are a series of tables, included and dimensioned according to the number of queues configured and options selected.

I strongly recommend that application code does not access these data structures directly but uses the provided API functions. This avoids incompatibility with future versions of Nucleus SE and unwanted side-effects and simplifies porting of an application to Nucleus RTOS. The details of data structures are included here to facilitate easier understanding of the working of the service call code and for debugging.

Kernel RAM data

These data structures are:

`NUSE_Queue_Head[]`—This is an array of type `U8`, with one entry for each configured queue, which represents a pointer to the front of the queue of messages. It is used as an index off of the addresses in `NUSE_Queue_Data[]` (see below).

`NUSE_Queue_Tail[]`—This is an array of type `U8`, with one entry for each configured queue, which represents a pointer to the end of the queue of messages. It is used as an index off of the addresses in `NUSE_Queue_Data[]` (see below).

`NUSE_Queue_Items[]`—This is an array of type `U8`, with one entry for each configured queue, which represents a count of the current number of messages in the queue. This data is arguably redundant, as its value can be derived from the head and tail indexes, but storing the count simplifies the code.

`NUSE_Queue_Blocking_Count[]`—This type `U8` array contains the counts of how many tasks are blocked on each queue. This array only exists if blocking API call support is enabled.

These data structures are all initialized to zeros by `NUSE_Init_Queue()` when Nucleus SE starts up. This is logical, as it renders every queue as being empty (unused). Chapter 17, *Nucleus SE Initialization and Start-up*, provides a full description of Nucleus SE start-up procedures.

Here are the definitions of these data structures in `nuse_init.c` file:

```
RAM U8 NUSE_Queue_Head[NUSE_QUEUE_NUMBER];
RAM U8 NUSE_Queue_Tail[NUSE_QUEUE_NUMBER];
RAM U8 NUSE_Queue_Items[NUSE_QUEUE_NUMBER];

#if NUSE_BLOCKING_ENABLE
  RAM U8 NUSE_Queue_Blocking_Count[NUSE_QUEUE_NUMBER];
#endif
```

User RAM

It is the user's responsibility to provide an area of RAM for data storage for each configured queue. The size of this RAM area must accommodate an array of type ADDR with one entry for each message in the queue.

ROM data

These data structures are:

NUSE_Queue_Data[]—This is an array of type ADDR, with one entry for each configured queue, which represents a pointer to the data area (discussed in "User RAM" sec-tion) for each queue.

NUSE_Queue_Size[]—This is an array of type U8, with one entry for each configured queue, which represents the num-ber of messages that may be accommodated by each queue.

These data structures are all declared and initialized (stati-cally, of course) in `nuse_config.c`; thus:

```
ROM ADDR *NUSE_Queue_Data[NUSE_QUEUE_NUMBER] =
{
  /* addresses of queue data areas ------ */
};

ROM U8 NUSE_Queue_Size[NUSE_QUEUE_NUMBER] =
{
  /* queue sizes ------ */
};
```

Queue data footprint

Like all kernel objects in Nucleus SE, the amount of data memory required for queues is readily predictable.

The ROM data footprint (in bytes) for all the queues in an application may be computed thus:

`NUSE_QUEUE_NUMBER * (sizeof(ADDR) + 1)`

The kernel RAM data footprint (in bytes) for all the queues in an application, when blocking API calls is enabled, may be computed thus:

`NUSE_QUEUE_NUMBER * 4`

Otherwise, it is:

`NUSE_QUEUE_NUMBER * 3`

The amount of user RAM (in bytes) required for the queue with index `queue` is:

`NUSE_Queue_Size[queue] * sizeof(ADDR)`

Unimplemented API calls

Four queue API calls found in Nucleus RTOS are not implemented in Nucleus SE.

Create queue

This API call creates a queue. It is not needed with Nucleus SE, as queues are created statically.

Service call prototype:

```
STATUS NU_Create_Queue(
NU_QUEUE *queue,
char *name,
VOID *start_address,
UNSIGNED queue_size,
OPTION message_type,
UNSIGNED message_size,
OPTION suspend_type);
```

Parameters:

queue—pointer to a user-supplied queue control block; this will be used as a "handle" for the queue in other API calls;

name—pointer to a 7-character, NULL-terminated name for the queue;

start_address—starting address for the queue;

message_type—type of message supported by the queue; may be NU_FIXED_SIZE or NU_VARIABLE_SIZE;

message_size—if the queue supports fixed-size messages, this parameter specifies the exact size of each message; otherwise, if the queue supports variable sized messages, this is the maximum message size;

suspend_type—specifies how tasks suspend on the queue. Valid options for this parameter are NU_FIFO and NU_PRIORITY, which represent FIFO and priority-order task suspension, respectively.

Returns:

NU_SUCCESS—indicates successful completion of the service.

NU_INVALID_QUEUE—indicates the queue control block pointer is NULL or already in use.

NU_INVALID_MEMORY—indicates the memory area specified by the start_address is invalid.

NU_INVALID_MESSAGE—indicates that the message_type parameter is invalid.

NU_INVALID_SIZE—indicates that either the message size is greater than the queue size, or that the queue size or message size is zero.

NU_INVALID_SUSPEND—indicates that the suspend_type parameter is invalid.

Delete queue

This API call deletes a previously created queue. It is not needed with Nucleus SE, as queues are created statically and cannot be deleted.

Service call prototype:

```
STATUS NU_Delete_Queue(NU_QUEUE *queue);
```

Parameters:

queue—pointer to queue control block

Returns:

NU_SUCCESS—indicates successful completion of the service.

NU_INVALID_QUEUE—indicates the queue pointer is invalid.

Queue pointers

This API call builds a sequential list of pointers to all queues in the system. It is not needed with Nucleus SE, as queues are identified by a simple index, not a pointer, and it would be redundant.

Service call prototype:

```
UNSIGNED NU_Queue_Pointers(
NU_QUEUE **pointer_list,
UNSIGNED maximum_pointers);
```

Parameters:

pointer_list—pointer to an array of NU_QUEUE pointers; this array will be filled with pointers to established queues in the system;

maximum_pointers—the maximum number of pointers to place in the array.

Returns:

The number of NU_QUEUE pointers placed into the array

Broadcast to queue

This API call broadcasts a message to all tasks waiting for a message from the specified queue. It is not implemented with Nucleus SE, as it would have added excessive complexity.

Service call prototype:

```
STATUS NU_Broadcast_To_Queue(
NU_QUEUE *queue,
VOID *message,
UNSIGNED size,
UNSIGNED suspend);
```

Parameters:

queue—pointer to queue control block;

message—pointer to the broadcast message;

size—the number of UNSIGNED data elements in the message. If the queue supports variable-length messages, this parameter must be equal to or less than the message size supported by the queue. If the queue supports fixed-size messages, this parameter must be exactly the same as the message size supported by the queue;

suspend—specifies whether or not to suspend the calling task if the queue is already full; valid options for this parameter are NU_NO_SUSPEND, NU_SUSPEND, or a timeout value.

Returns:

NU_SUCCESS—indicates successful completion of the service.

NU_INVALID_QUEUE—indicates the queue pointer is invalid.

NU_INVALID_POINTER—indicates that the message pointer is NULL.

NU_INVALID_SIZE—indicates that the message size speci-fied is not compatible with the size specified when the queue was created.

NU_INVALID_SUSPEND—indicates that suspend attempted from a nontask thread.

NU_QUEUE_FULL—indicates that there is insufficient space in the queue for the message.

NU_TIMEOUT—indicates the queue is still full after the timeout has expired.

NU_QUEUE_DELETED—Queue was deleted, while the task was suspended.

NU_QUEUE_RESET—Queue was reset, while the task was suspended.

Compatibility with Nucleus RTOS

With all aspects of Nucleus SE, it was my goal to maintain as high a level of applications code compatibility with Nucleus RTOS as possible. Queues are no exception and, from a user's perspective, they are implemented in much the same way as in Nucleus RTOS. There are areas of incompatibility, which have come about where I determined that such an incompatibility would be acceptable, given that the resulting code is easier to understand, or, more likely, could be made more memory efficient. Otherwise, Nucleus RTOS API calls may be almost directly mapped onto Nucleus SE calls. Chapter 20, *Using Nucleus SE*, includes further information on using Nucleus SE for users of Nucleus RTOS.

Object identifiers

In Nucleus RTOS, all objects are described by a data struc-ture—a control block—which has a specific data type. A pointer to this control block serves as an identifier for the queue. In Nucleus SE, I decided that a different approach was needed for memory efficiency, and all kernel objects are described by a number of tables in RAM and/or ROM. The size of these tables is determined by the number of each object type that is configured. The identifier for a specific object is simply an index into those tables. So, I have defined **NUSE_QUEUE** as being equivalent to **U8**; a variable—not a pointer—of this type then serves as the queue identifier. This is a small incompatibility, which is easily handled if code is ported to or from Nucleus RTOS. Object identifiers are normally just stored and passed around and not operated on in any way.

Nucleus RTOS also supports naming of queues. These names are only used for target-based debug facilities. I omitted them from Nucleus SE to save memory.

Message size and variability

In Nucleus RTOS, a queue may be configured to handle messages that are comprised of any number of **unsigned** data elements. In Nucleus SE, queues are simplified and only single **ADDR** messages are supported. Pipes are a little more flexible in Nucleus SE and may offer a useful alternative to queues for some applications; pipes will be covered in the next chapter.

Nucleus RTOS also supports queues with variable size messages, where only the maximum size is specified at creation time. Variable size messages are not supported by Nucleus SE.

Queue size

The number of messages in a queue in Nucleus SE is limited to 256, as all the index variables and constants are type **U8**. Nucleus RTOS is not limited in this way.

Unimplemented API calls

Nucleus RTOS supports 10 service calls to work with queues. Of these, four are not implemented in Nucleus SE. Details of these and of the decision to omit them may be found in *"Unimplemented API calls"* section earlier in this chapter.

13

Pipes

Pipes were introduced in Chapter 3, *RTOS Services and Facilities*. They provide a more flexible means of passing simple messages between tasks than mailboxes or queues.

Using pipes

In Nucleus SE, pipes are configured at build time. There may be a maximum of 16 pipes configured for an application. If no pipes are configured, no data structures or service call code appertaining to pipes are included in the application.

A pipe is simply a set of storage locations, each big enough to hold a single data item of user-defined byte length, access to which is controlled so that it may be safely utilized by multiple tasks. Tasks can write to a pipe repeatedly until all the locations are full. Tasks can read from a pipe and data is normally received on a first-in, first-out (FIFO) basis. Trying to send to a full pipe or read from an empty one may result in an error or task suspension, depending on options selected in the application program interface (API) call and the Nucleus SE configuration.

Pipes and queues

Nucleus SE also supports queues, which were covered in detail in the last chapter. The main difference between pipes and queues is the message size. Queues carry messages comprising a single ADDR—these would commonly be pointers. A pipe carries messages that are an arbitrary number of bytes long; the size is fixed for each pipe in the application and set at configuration time.

Configuring pipes

Number of pipes

As with most aspects of Nucleus SE, the configuration of pipes is primarily controlled by #define statements in nuse_config.h. The

Embedded RTOS Design. DOI: https://doi.org/10.1016/B978-0-12-822851-7.00013-8

key setting is NUSE_PIPE_NUMBER, which determines how many pipes are configured for the application. The default setting is 0 (i.e., no pipes are in use) and you can set it to any value up to 16. An erroneous value will result in a compile time error, which is generated by a test in nuse_config_check.h (this is included into nuse_config.c and hence compiled with this module), resulting in a #error statement being compiled.

Choosing a nonzero value is the "master enable" for pipes. This results in some data structures being defined and sized accordingly, of which more later in this chapter. It also activates the API enabling settings.

API enables

Every API function (service call) in Nucleus SE has an enabling #define symbol in nuse_config.h. For pipes, these are:

```
NUSE_PIPE_SEND
NUSE_PIPE_RECEIVE
NUSE_PIPE_JAM
NUSE_PIPE_RESET
NUSE_PIPE_INFORMATION
NUSE_PIPE_COUNT
```

By default, all of these are set to FALSE, thus disabling each service call and inhibiting the inclusion of any implementation code. To configure pipes for an application, you need to select the API calls that you want to use and set their enabling symbols to TRUE.

Here is an extract from the default nuse_config.h file.

```
#define NUSE_PIPE_NUMBER    0   /* Number of pipes in the
                                   system - 0-16 */
                    /* Service call enablers */
#define NUSE_PIPE_SEND  FALSE
#define NUSE_PIPE_RECEIVE  FALSE
#define NUSE_PIPE_JAM  FALSE
#define NUSE_PIPE_RESET  FALSE
#define NUSE_PIPE_INFORMATION  FALSE
#define NUSE_PIPE_COUNT  FALSE
```

A compile time error will result if a pipe API function is enabled and no pipes are configured (except for NUSE_Pipe_Count(), which is always permitted). If your code uses an API call, which has not been enabled, a link time error will result, as no implementation code will have been included in the application.

Pipe service calls

Nucleus RTOS supports 10 service calls that appertain to pipes, which provide the following functionality:

Send a message to a pipe. Implemented by `NUSE_Pipe_Send()` in Nucleus SE.

Receive a message from a pipe. Implemented by `NUSE_Pipe_Receive()` in Nucleus SE.

Send a message to the front of a pipe. Implemented by `NUSE_Pipe_Jam()` in Nucleus SE.

Restore a pipe to the unused state, with no tasks suspended (reset). Implemented by `NUSE_Pipe_Reset()` in Nucleus SE.

Provide information about a specified pipe. Implemented by `NUSE_Pipe_Information()` in Nucleus SE.

Return a count of how many pipes are (currently) configured for the application. Implemented by `NUSE_Pipe_Count()` in Nucleus SE.

Add a new pipe to the application (create). Not implemented in Nucleus SE.

Remove a pipe from the application (delete). Not implemented in Nucleus SE.

Return pointers to all the pipes (currently) in the application. Not implemented in Nucleus SE.

Send a message to all the tasks that are suspended on a pipe (broadcast). Not implemented in Nucleus SE.

The implementation of each of these service calls will be examined in detail.

Pipe write and read services

The fundamental operations, which can be performed on a pipe, are writing data to it—which is sometimes termed *sending*—and reading data from it—which is also termed *receiving*. It is also possible to write data to the front of a pipe—which is also termed *jamming*. Nucleus RTOS and Nucleus SE each provide three basic API calls for these operations, which will be discussed here.

Writing to a pipe

The Nucleus RTOS API call for writing to a pipe is very flexible, enabling you to suspend indefinitely, or with a timeout, if the operation cannot be completed immediately; that is, you try to write to a full pipe. Nucleus SE provides the same service, except task suspend is optional and timeout is not implemented.

Nucleus RTOS also offers a facility to broadcast to a pipe, but this is not supported by Nucleus SE. It will be described under the *"Unimplemented API calls"* section later in this chapter.

Nucleus RTOS API call for sending to a pipe

Service call prototype:

```
STATUS NU_Send_To_Pipe(
NU_PIPE *pipe,
VOID *message,
UNSIGNED size,
UNSIGNED suspend);
```

Parameters:

pipe—pointer to the user-supplied pipe control block;

message—a pointer to the message to be sent;

size—the number of bytes in the message. If the pipe supports variable-length messages, this parameter must be equal to or less than the message size supported by the pipe. If the pipe supports fixed-size messages, this parameter must be exactly the same as the message size supported by the pipe;

suspend—specification for task suspend; may be NU_NO_SUSPEND or NU_SUSPEND or a timeout value.

Returns:

NU_SUCCESS—The call was completed successfully.

NU_INVALID_PIPE—The pipe pointer is invalid.

NU_INVALID_POINTER—The message pointer is NULL.

NU_INVALID_SIZE—The message size is incompatible with the message size supported by the pipe.

NU_INVALID_SUSPEND—Suspend was attempted from a nontask thread.

NU_PIPE_FULL—The pipe is full and suspend was not specified.

NU_TIMEOUT—The pipe is still full even after suspending for the specified timeout value.

NU_PIPE_DELETED—The pipe was deleted, while the task was suspended.

NU_PIPE_RESET—The pipe was reset, while the task was suspended.

Nucleus SE API Call for sending to a pipe

This API call supports the key functionality of the Nucleus RTOS API.

Service call prototype:

```
STATUS NUSE_Pipe_Send(
NUSE_PIPE pipe,
U8 *message,
U8 suspend);
```

Parameters:

pipe—the index (ID) of the pipe to be utilized;

message—a pointer to the message to be sent, which is a sequence of bytes as long as the configured message size of the pipe;

suspend—specification for task suspend; may be **NUSE_NO_SUSPEND** or **NUSE_SUSPEND**.

Returns:

NUSE_SUCCESS—The call was completed successfully.

NUSE_INVALID_PIPE—The pipe index is invalid.

NUSE_INVALID_POINTER—The message pointer is **NULL**.

NUSE_INVALID_SUSPEND—Suspend was attempted from a non-task thread or when blocking API calls were not enabled.

NUSE_PIPE_FULL—The pipe is full and suspend was not specified.

NUSE_PIPE_WAS_RESET—The pipe was reset, while the task was suspended.

Nucleus SE implementation of pipe send

The bulk of the code of the `NUSE_Pipe_Send()` API function—after parameter checking—is selected by conditional compilation, dependent on whether support for blocking (task suspend) API calls is enabled. We will look at the two variants separately here.

If blocking is not enabled, the code for this API call is quite simple:

```
if (NUSE_Pipe_Items[pipe] == NUSE_Pipe_Size[pipe]) /* pipe full */
{
  return_value = NUSE_PIPE_FULL;
}
else                    /* pipe element available */
{
  data = &NUSE_Pipe_Data[pipe][NUSE_Pipe_Head[pipe]];
  for (i = 0; i < msgsize; i++)
  {
    *data++ = *message++;
  }
  NUSE_Pipe_Head[pipe] + = msgsize;
  if (NUSE_Pipe_Head[pipe] == (NUSE_Pipe_Size[pipe] * msgsize))
  {
    NUSE_Pipe_Head[pipe] = 0;
  }
  NUSE_Pipe_Items[pipe]++;
  return_value = NUSE_SUCCESS;
}
```

The function simply checks that there is room in the pipe and uses the `NUSE_Pipe_Head[]` index to store the message in the pipe's data area.

When blocking is enabled, the code becomes more complex:

```
do
{
  if (NUSE_Pipe_Items[pipe] == NUSE_Pipe_Size[pipe]) /* pipe
                                                        full */
  {
    if (suspend == NUSE_NO_SUSPEND)
    {
      return_value = NUSE_PIPE_FULL;
    }
    else                        /* block task */
    {
      NUSE_Pipe_Blocking_Count[pipe]++;
      NUSE_Suspend_Task(NUSE_Task_Active, (pipe << 4) |
                            NUSE_PIPE_SUSPEND);
```

```
        return_value =
        NUSE_Task_Blocking_Return[NUSE_Task_Active];
        if (return_value! = NUSE_SUCCESS)
        {
          suspend = NUSE_NO_SUSPEND;
        }
      }
    }
    else              /* pipe element available */
    {
      data = &NUSE_Pipe_Data[pipe][NUSE_Pipe_Head[pipe]];
      for (i = 0; i < msgsize; i++)
      {
        *data++ = *message++;
      }
      NUSE_Pipe_Head[pipe] + =msgsize;
      if (NUSE_Pipe_Head[pipe] == (NUSE_Pipe_Size[pipe] *
                                              msgsize))
      {
        NUSE_Pipe_Head[pipe] = 0;
      }
      NUSE_Pipe_Items[pipe]++;
      if (NUSE_Pipe_Blocking_Count[pipe]! = 0)
      {
        U8 index;   /* check whether a task is blocked on this pipe */

        NUSE_Pipe_Blocking_Count[pipe]--;
        for (index = 0; index < NUSE_TASK_NUMBER; index++)
        {
          if ((LONIB(NUSE_Task_Status[index]) ==
          NUSE_PIPE_SUSPEND)
              && (HINIB(NUSE_Task_Status[index]) ==pipe))
          {
            NUSE_Task_Blocking_Return[index] = NUSE_SUCCESS;
            NUSE_Wake_Task(index);
            break;
          }
        }
      }
      return_value = NUSE_SUCCESS;
      suspend = NUSE_NO_SUSPEND;
    }
  } while (suspend == NUSE_SUSPEND);
```

Some explanation of the code may be useful:

The code is enclosed in a do…while loop, which continues while the parameter suspend has the value NUSE_SUSPEND.

If the pipe is full and suspend is set to NUSE_NO_SUSPEND, the API call exits with NUSE_PIPE_FULL. If suspend was set to

NUSE_SUSPEND, the task is suspended. On return (i.e., when the task is woken up), if the return value is NUSE_SUCCESS, indicating that the task was woken because a message had been read (as opposed to a reset of the pipe), the code loops back to the top.

If the pipe is not full, the supplied message is stored using the NUSE_Pipe_Head[] index to store the message in the pipe's data area. A check is made on whether any tasks are suspended (waiting to receive) on the pipe. If there are any tasks waiting, the first one is woken. The suspend variable is set to NUSE_NO_SUSPEND and the API call exits with NUSE_SUCCESS.

Reading from a pipe

The Nucleus RTOS API call for reading from a pipe is very flexible, enabling you to suspend indefinitely, or with a timeout, if the operation cannot be completed immediately; that is, you try to read from an empty pipe. Nucleus SE provides the same service, except task suspend is optional and timeout is not implemented.

Nucleus RTOS API call for receiving from a pipe

Service call prototype:

```
STATUS NU_Receive_From_Pipe(
NU_PIPE *pipe,
VOID *message,
UNSIGNED size,
UNSIGNED *actual_size,
UNSIGNED suspend);
```

Parameters:

pipe—pointer to user-supplied pipe control block;

message—pointer to storage for the message to be received;

size—the number of bytes in the message. This number must correspond to the message size defined when the pipe was created;

suspend—specification for task suspend; may be NU_NO_SUSPEND or NU_SUSPEND or a timeout value;

Returns:

NU_SUCCESS—The call was completed successfully.

NU_INVALID_PIPE—The pipe pointer is invalid.

NU_INVALID_POINTER—The message pointer is NULL.

NU_INVALID_SUSPEND—Suspend was attempted from a nontask thread.

NU_PIPE_EMPTY—The pipe is empty and suspend was not specified.

NU_TIMEOUT—indicates that the pipe is still empty even after suspending for the specified timeout value.

NU_PIPE_DELETED—The pipe was deleted, while the task was suspended.

NU_PIPE_RESET—The pipe was reset, while the task was suspended.

Nucleus SE API call for receiving from a pipe

This API call supports the key functionality of the Nucleus RTOS API.

Service call prototype:

```
STATUS NUSE_Pipe_Receive(
NUSE_PIPE pipe,
U8 *message,
U8 suspend);
```

Parameters:

pipe—the index (ID) of the pipe to be utilized;

message—a pointer to storage for the message to be received, which is a sequence of bytes as long as the configured message size of the pipe;

suspend—specification for task suspend; may be NUSE_NO_SUSPEND or NUSE_SUSPEND.

Returns:

NUSE_SUCCESS—The call was completed successfully.

NUSE_INVALID_PIPE—The pipe index is invalid.

NUSE_INVALID_POINTER—The message pointer is **NULL**.

NUSE_INVALID_SUSPEND—Suspend was attempted from a non-task thread or when blocking API calls were not enabled.

NUSE_PIPE_EMPTY—The pipe is empty and suspend was not specified.

NUSE_PIPE_WAS_RESET—The pipe was reset, while the task was suspended.

Nucleus SE implementation of pipe receive

The bulk of the code of the **NUSE_Pipe_Receive()** API function—after parameter checking—is selected by conditional compilation, dependent on whether support for blocking (task suspend) API calls is enabled. We will look at the two variants separately here.

If blocking is not enabled, the code for this API call is quite simple:

```
if (NUSE_Pipe_Items[pipe] == 0)          /* pipe empty */
{
  return_value = NUSE_PIPE_EMPTY;
}
else
{               /* message available */
  data = &NUSE_Pipe_Data[pipe][NUSE_Pipe_Tail[pipe]];
  for (i = 0; i < msgsize; i++)
  {
    *message++ = *data++;
  }

  NUSE_Pipe_Tail[pipe] += msgsize;
  if (NUSE_Pipe_Tail[pipe] == (NUSE_Pipe_Size[pipe] * msgsize))
  {
    NUSE_Pipe_Tail[pipe] = 0;
  }

  NUSE_Pipe_Items[pipe]--;

  *actual_size = msgsize;
  return_value = NUSE_SUCCESS;
}
```

The function simply checks that there is a message in the pipe and uses the **NUSE_Pipe_Tail[]** index to obtain the message from the pipe's data area and returns the data via the message pointer.

When blocking is enabled, the code becomes more complex:

```
do
{
  if (NUSE_Pipe_Items[pipe] == 0)     /* pipe empty */
  {
    if (suspend == NUSE_NO_SUSPEND)
    {
      return_value = NUSE_PIPE_EMPTY;
    }
    else
    {                      /* block task */
      NUSE_Pipe_Blocking_Count[pipe]++;
      NUSE_Suspend_Task(NUSE_Task_Active, (pipe << 4) |
                                 NUSE_PIPE_SUSPEND);
      return_value =
      NUSE_Task_Blocking_Return[NUSE_Task_Active];
      if (return_value! = NUSE_SUCCESS)
      {
        suspend = NUSE_NO_SUSPEND;
      }
    }
  }
  else
  {                  /* message available */
    data = &NUSE_Pipe_Data[pipe][NUSE_Pipe_Tail[pipe]];
    for (i = 0; i < msgsize; i++)
    {
      *message++ = *data++;
    }
    NUSE_Pipe_Tail[pipe] + = msgsize;
    if (NUSE_Pipe_Tail[pipe] == (NUSE_Pipe_Size[pipe] * msgsize))
    {
      NUSE_Pipe_Tail[pipe] = 0;
    }
    NUSE_Pipe_Items[pipe]--;
    if (NUSE_Pipe_Blocking_Count[pipe]! = 0)
    {
      U8 index;        /* check whether a task is blocked */
                /* on this pipe */

      NUSE_Pipe_Blocking_Count[pipe]--;
      for (index = 0; index < NUSE_TASK_NUMBER; index++)
      {
```

```
            if ((LONIB(NUSE_Task_Status[index]) ==
                            NUSE_PIPE_SUSPEND)
                && (HINIB(NUSE_Task_Status[index]) ==pipe))
            {
              NUSE_Task_Blocking_Return[index] = NUSE_SUCCESS;
              NUSE_Wake_Task(index);
              break;
            }
          }
        }
        *actual_size = msgsize;
        return_value = NUSE_SUCCESS;
        suspend = NUSE_NO_SUSPEND;
      }
  } while (suspend == NUSE_SUSPEND);
```

Some explanation of the code may be useful:

The code is enclosed in a do...while loop, which continues while the parameter suspend has the value NUSE_SUSPEND.

If the pipe is empty and suspend is set to NUSE_NO_SUSPEND, the API call exits with NUSE_PIPE_EMPTY. If suspend was set to NUSE_SUSPEND, the task is suspended. On return (i.e., when the task is woken up), if the return value is NUSE_SUCCESS, indicating that the task was woken because a message had been sent (as opposed to a reset of the pipe), the code loops back to the top.

If the pipe contains any messages, a stored message is returned using the NUSE_Pipe_Tail[] index to obtain the message from the pipe's data area. A check is made on whether any tasks are suspended (waiting to send) on the pipe. If there are any tasks waiting, the first one is woken. The suspend variable is set to NUSE_NO_SUSPEND and the API call exits with NUSE_SUCCESS.

Writing to the front of a pipe

The Nucleus RTOS API call for writing to the front of a pipe is very flexible, enabling you to suspend indefinitely, or with a timeout, if the operation cannot be completed immediately; that is, you try to write to a full pipe. Nucleus SE provides the same service, except task suspend is optional and timeout is not implemented.

Nucleus RTOS API call for jamming to a pipe

Service call prototype:

```
STATUS NU_Send_To_Front_Of_Pipe(
NU_PIPE *pipe,
VOID *message,
UNSIGNED size,
UNSIGNED suspend);
```

Parameters:

pipe—pointer to a user-supplied pipe control block;

message—a pointer to the message to be sent;

size—the number of bytes in the message. If the pipe supports variable-length messages, this parameter must be equal to or less than the message size supported by the pipe. If the pipe supports fixed-size messages, this parameter must be exactly the same as the message size supported by the pipe;

suspend—specification for task suspend; may be NU_NO_SUSPEND or NU_SUSPEND or a timeout value.

Returns:

NU_SUCCESS—The call was completed successfully.

NU_INVALID_PIPE—The pipe pointer is invalid.

NU_INVALID_POINTER—The message pointer is NULL.

NU_INVALID_SIZE—The message size is incompatible with the message size supported by the pipe.

NU_INVALID_SUSPEND—Suspend was attempted from a nontask thread.

NU_PIPE_FULL—The pipe is full and suspend was not specified.

NU_TIMEOUT—The pipe is still full even after suspending for the specified timeout value.

NU_PIPE_DELETED—The pipe was deleted, while the task was suspended.

NU_PIPE_RESET—The pipe was reset, while the task was suspended.

Nucleus SE API call for jamming to a pipe

This API call supports the key functionality of the Nucleus RTOS API.

Service call prototype:

```
STATUS NUSE_Pipe_Jam(
NUSE_PIPE pipe,
ADDR *message,
U8 suspend);
```

Parameters:

pipe—the index (ID) of the pipe to be utilized;

message—a pointer to the message to be sent, which is a sequence of bytes as long as the configured message size of the pipe;

suspend—specification for task suspend; may be NUSE_NO_SUSPEND or NUSE_SUSPEND.

Returns:

NUSE_SUCCESS—The call was completed successfully.

NUSE_INVALID_PIPE—The pipe index is invalid.

NUSE_INVALID_POINTER—The message pointer is NULL.

NUSE_INVALID_SUSPEND—Suspend was attempted from a non-task thread or when blocking API calls were not enabled.

NUSE_PIPE_FULL—The pipe is full and suspend was not specified.

NUSE_PIPE_WAS_RESET—The pipe was reset, while the task was suspended.

Nucleus SE implementation of jamming to a pipe

The bulk of the code of the NUSE_Pipe_Jam() API function is very similar to that of NUSE_Pipe_Send(), except that the data are stored using the NUSE_Pipe_Tail[] index; thus:

```
if (NUSE_Pipe_Items[pipe] == NUSE_Pipe_Size[pipe]) /* pipe full */
{
  return_value = NUSE_PIPE_FULL;
}
else                    /* pipe element available */
{
  if (NUSE_Pipe_Tail[pipe] == 0)
  {
    NUSE_Pipe_Tail[pipe] = (NUSE_Pipe_Size[pipe] - 1) * msgsize;
```

```
    }
    else
    {
      NUSE_Pipe_Tail[pipe]-=msgsize;
    }
    data=&NUSE_Pipe_Data[pipe][NUSE_Pipe_Tail[pipe]];
    for (i=0; i<msgsize; i++)
    {
      *data++=*message++;
    }
    NUSE_Pipe_Items[pipe]++;
    return_value=NUSE_SUCCESS;
  }
```

Pipe utility services

Nucleus RTOS has four API calls that provide utility functions associated with pipes: reset pipe, return information about a pipe, return number of pipes in the application, and return pointers to all pipes in the application. The first three of these are implemented in Nucleus SE.

Resetting a pipe

This API call restores the pipe to its initial, unused state. Any messages stored in the pipe are lost. Any tasks that were suspended on the pipe are resumed and receive a return code of NUSE_PIPE_WAS_RESET.

Nucleus RTOS API call for resetting a pipe

Service call prototype:

    STATUS NU_Reset_Pipe(NU_PIPE *pipe);

Parameters:

pipe—pointer to user-defined pipe control block

Returns:

NU_SUCCESS—The call was completed successfully.

NU_INVALID_PIPE—The pipe pointer is not valid.

Nucleus SE API call for resetting a pipe

This API call supports the key functionality of the Nucleus RTOS API.

Service call prototype:

```
STATUS NUSE_Pipe_Reset(NUSE_PIPE pipe);
```

Parameters:

pipe—the index (ID) of the pipe to be reset

Returns:

NUSE_SUCCESS—The call was completed successfully.

NUSE_INVALID_PIPE—The pipe index is not valid.

Nucleus SE implementation of pipe reset

The initial part of the code of the **NUSE_Pipe_Reset()** API function—after parameter checking—is quite straightforward. The head and tail indexes and the pipe's message count are all set to zero.

When blocking is enabled, additional code takes care of waking up any suspended tasks; thus:

```
while (NUSE_Pipe_Blocking_Count[pipe] != 0)
{
  U8 index; /* check whether any tasks are blocked on this pipe */

  for (index = 0; index < NUSE_TASK_NUMBER; index++)
  {
    if ((LONIB(NUSE_Task_Status[index]) ==
                      NUSE_PIPE_SUSPEND)
      && (HINIB(NUSE_Task_Status[index]) == pipe))
    {
      NUSE_Task_Blocking_Return[index] = NUSE_PIPE_RESET;
      NUSE_Task_Status[index] = NUSE_READY;
      break;
    }
  }
  NUSE_Pipe_Blocking_Count[pipe]--;
}

#if NUSE_SCHEDULER_TYPE == NUSE_PRIORITY_SCHEDULER
  NUSE_Reschedule(NUSE_NO_TASK);
#endif
```

Each task suspended on the pipe is marked as *Ready* with a suspend return code of NUSE_PIPE_WAS_RESET. After this process is complete, if the Priority scheduler is in use, a call is made to NUSE_Reschedule(), as one or more higher priority tasks may have been readied and needs to be allowed to run.

Pipe information

This service call obtains a selection of information about a pipe. The Nucleus SE implementation differs from Nucleus RTOS in that it returns less information, as object naming, variable message size, and suspend ordering are not supported and task suspend may not be enabled.

Nucleus RTOS API call for pipe information

Service call prototype:

```
STATUS NU_Pipe_Information(
NU_PIPE *pipe,
CHAR *name,
VOID **start_address,
UNSIGNED *pipe_size,
UNSIGNED *available,
UNSIGNED *messages,
OPTION *message_type,
UNSIGNED *message_size,
OPTION *suspend_type,
UNSIGNED *tasks_waiting,
NU_TASK **first_task);
```

Parameters:

pipe—pointer to the user-supplied pipe control block;

name—pointer to an 8-character destination area for the message-pipe's name;

start_address—a pointer to a pointer, which will receive the address of the start of the pipe's data area;

pipe_size—a pointer to a variable for holding the total number of bytes in the pipe;

available—a pointer to a variable for holding the number of available bytes in the pipe;

messages—a pointer to a variable for holding the number of messages currently in the pipe;

message_type—pointer to a variable for holding the type of messages supported by the pipe; valid message types are NU_FIXED_SIZE and NU_VARIABLE_SIZE;

message_size—pointer to a variable for holding the number of bytes in each pipe message; if the pipe supports variable-length messages, this number is the maximum message size; suspend_type—pointer to a variable for holding the task suspend type. Valid task suspend types are NU_FIFO and NU_PRIORITY;

tasks_waiting—a pointer to a variable, which will receive the number of tasks suspended on this pipe;

first_task—a pointer to a task pointer; the pointer of the first suspended task is placed in this task pointer.

Returns:

NU_SUCCESS—The call was completed successfully.

NU_INVALID_PIPE—The pipe pointer is not valid.

Nucleus SE API call for pipe information

This API call supports the key functionality of the Nucleus RTOS API.

Service call prototype:

```
STATUS NUSE_Pipe_Information(
NUSE_PIPE pipe,
ADDR *start_address,
U8 *pipe_size,
U8 *available,
U8 *messages,
U8 *message_size,
U8 *tasks_waiting,
NUSE_TASK *first_task);
```

Parameters:

pipe—the index of the pipe about which information is being requested;

start_address—a pointer to a variable of type ADDR, which will receive the address of the start of the pipe's data area;

pipe_size—a pointer to a variable of type U8, which will receive the total number of messages for which the pipe has capacity;

available—a pointer to a variable of type **U8**, which will receive the number of messages for which the pipe has currently remaining capacity;

messages—a pointer to a variable of type **U8**, which will receive the number of messages currently in the pipe;

message_size—a pointer to a variable of type **U8**, which will receive the size of messages handled by this pipe;

tasks_waiting—a pointer to a variable, which will receive the number of tasks suspended on this pipe (nothing returned if task suspend is disabled);

first_task—a pointer to a variable of type **NUSE_TASK**, which will receive the index of the first suspended task (nothing returned if task suspend is disabled).

Returns:

NUSE_SUCCESS—The call was completed successfully.

NUSE_INVALID_PIPE—The pipe index is not valid.

NUSE_INVALID_POINTER—One or more of the pointer parameters is invalid.

Nucleus SE implementation of pipe information

The implementation of this API call is quite straightforward:

```
*start_address = NUSE_Pipe_Data[pipe];
*pipe_size = NUSE_Pipe_Size[pipe];
*available = NUSE_Pipe_Size[pipe] - NUSE_Pipe_Items[pipe];
*messages = NUSE_Pipe_Items[pipe];
*message_size = NUSE_Pipe_Message_Size[pipe];

#if NUSE_BLOCKING_ENABLE
  *tasks_waiting = NUSE_Pipe_Blocking_Count[pipe];
  if (NUSE_Pipe_Blocking_Count[pipe] != 0)
  {
    U8 index;

    for (index = 0; index < NUSE_TASK_NUMBER; index++)
    {
      if ((LONIB(NUSE_Task_Status[index]) ==
                        NUSE_PIPE_SUSPEND)
          && (HINIB(NUSE_Task_Status[index]) == pipe))
      {
```

```
        *first_task = index;
        break;
      }
    }
  }
  else
  {
    *first_task = 0;
  }
#else
  *tasks_waiting = 0;
  *first_task = 0;
#endif
```

The function returns the pipe status. Then, if blocking API calls is enabled, the number of waiting tasks and the index of the first one are returned (otherwise, these two parameters are set to 0).

Obtaining the number of pipes

This service call returns the number of pipes configured in the application. Whilst in Nucleus RTOS this will vary over time and the returned value will represent the current number of pipes, in Nucleus SE the value returned is set at build time and cannot change.

Nucleus RTOS API call for pipe count

Service call prototype:

```
UNSIGNED NU_Established_Pipes(VOID);
```

Parameters:

None

Returns:

The number of created pipes in the system.

Nucleus SE API call for pipe count

This API call supports the key functionality of the Nucleus RTOS API.

Service call prototype:

```
U8 NUSE_Pipe_Count(void);
```

Parameters:

None

Returns:

The number of configured pipes in the application

Nucleus SE implementation of pipe count

The implementation of this API call is almost trivially simple: the value of the `#define` symbol `NUSE_PIPE_NUMBER` is returned.

Data structures

Pipes utilize six or seven data structures—all in RAM and ROM—which, like other Nucleus SE objects, are a series of tables, included and dimensioned according to the number of pipes configured and options selected.

I strongly recommend that application code does not access these data structures directly but uses the provided API functions. This avoids incompatibility with future versions of Nucleus SE and unwanted side effects and simplifies porting of an application to Nucleus RTOS. The details of data structures are included here to facilitate easier understanding of the working of the service call code and for debugging.

Kernel RAM data

These data structures are:

`NUSE_Pipe_Head[]`—This is an array of type `U8`, with one entry for each configured pipe, which represents a pointer to the front of the pipe of messages. It is used as an index off of the addresses in `NUSE_Pipe_Data[]` (see below).

`NUSE_Pipe_Tail[]`—This is an array of type `U8`, with one entry for each configured pipe, which represents a pointer to the end of the pipe of messages. It is used as an index off of the addresses in `NUSE_Pipe_Data[]` (see below).

NUSE_Pipe_Items[]—This is an array of type **U8**, with one entry for each configured pipe, which represents a count of the current number of messages in the pipe. This data is arguably redundant, as its value can be derived from the head and tail indexes, but storing the count simplifies the code.

NUSE_Pipe_Blocking_Count[]—This type **U8** array contains the counts of how many tasks are blocked on each pipe. This array only exists if blocking API call support is enabled.

These data structures are all initialized to zeros by **NUSE_Init_Pipe()** when Nucleus SE starts up. This is logical, as it renders every pipe as being empty (unused). Chapter 17, *Nucleus SE Initialization and Start-up*, provides a full description of Nucleus SE start-up procedures.

Here are the definitions of these data structures in **nuse_init.c** file:

```
RAM U8 NUSE_Pipe_Head[NUSE_PIPE_NUMBER];
RAM U8 NUSE_Pipe_Tail[NUSE_PIPE_NUMBER];
RAM U8 NUSE_Pipe_Items[NUSE_PIPE_NUMBER];

#if NUSE_BLOCKING_ENABLE
  RAM U8 NUSE_Pipe_Blocking_Count[NUSE_PIPE_NUMBER];
#endif
```

User RAM

It is the user's responsibility to provide an area of RAM for data storage for each configured pipe. The size of this RAM area must accommodate an array of type **U8** large enough to accommodate all of the messages in the pipe.

ROM data

These data structures are:

NUSE_Pipe_Data[]—This is an array of type **ADDR**, with one entry for each configured pipe, which represents a pointer to the data area (discussed in "User RAM" section) for each pipe.

NUSE_Pipe_Size[]—This is an array of type **U8**, with one entry for each configured pipe, which represents the number of messages that may be accommodated by each pipe.

`NUSE_Pipe_Message_Size[]`—This is an array of type **u8**, with one entry for each configured pipe, which represents the size of messages (in bytes) that may be accommodated by each pipe.

These data structures are all declared and initialized (statically, of course) in **nuse_config.c**; thus:

```
ROM ADDR *NUSE_Pipe_Data[NUSE_PIPE_NUMBER] =
{
    /* addresses of pipe data areas ------ */
};

ROM U8 NUSE_Pipe_Size[NUSE_PIPE_NUMBER] =
{
    /* pipe sizes ------ */
};

ROM U8 NUSE_Pipe_Message_Size[NUSE_PIPE_NUMBER] =
{
    /* pipe message sizes ------ */
};
```

Pipe data footprint

Like all kernel objects in Nucleus SE, the amount of data memory required for pipes is readily predictable.

The ROM data footprint (in bytes) for all the pipes in an application may be computed thus:

```
NUSE_PIPE_NUMBER * (sizeof(ADDR) + 2)
```

The kernel RAM data footprint (in bytes) for all the pipes in an application, when blocking API calls is enabled, may be computed thus:

```
NUSE_PIPE_NUMBER * 4
```

Otherwise, it is:

```
NUSE_PIPE_NUMBER * 3
```

The amount of user RAM (in bytes) required for the pipe with index pipe is:

```
NUSE_Pipe_Size[pipe] * NUSE_Pipe_Message_Size[pipe]
```

Unimplemented API calls

Four pipe API calls found in Nucleus RTOS are not implemented in Nucleus SE.

Create pipe

This API call creates a pipe. It is not needed with Nucleus SE, as pipes are created statically.

Service call prototype:

```
STATUS NU_Create_Pipe(
NU_PIPE *pipe,
char *name,
VOID *start_address,
UNSIGNED pipe_size,
OPTION message_type,
UNSIGNED message_size,
OPTION suspend_type);
```

Parameters:

> `pipe`—pointer to a user-supplied pipe control block; this will be used as a "handle" for the pipe in other API calls;

> `name`—pointers to a 7-character, `NULL`-terminated name for the pipe;

> `start_address`—starting address for the pipe;

> `pipe_size`—the total number of bytes in the pipe;

> `message_type`—type of message supported by the pipe; may be `NU_FIXED_SIZE` or `NU_VARIABLE_SIZE`;

> `message_size`—if the pipe supports fixed-size messages, this parameter specifies the exact size of each message; otherwise, if the pipe supports variable sized messages, this is the maximum message size;

> `suspend_type`—specifies how tasks suspend on the pipe. Valid options for this parameter are `NU_FIFO` and `NU_PRIORITY`, which represent FIFO and priority-order task suspension, respectively.

Returns:

> `NU_SUCCESS`—indicates successful completion of the service.

NU_INVALID_PIPE—indicates the pipe control block pointer is NULL or already in use.

NU_INVALID_MEMORY—indicates the memory area specified by the start_address is invalid.

NU_INVALID_MESSAGE—indicates that the message_type parameter is invalid.

NU_INVALID_SIZE—indicates that either the message size is greater than the pipe size, or that the pipe size or message size is zero.

NU_INVALID_SUSPEND—indicates that the suspend_type parameter is invalid.

Delete pipe

This API call deletes a previously created pipe. It is not needed with Nucleus SE, as pipes are created statically and cannot be deleted. Service call prototype:

```
STATUS NU_Delete_Pipe(NU_PIPE *pipe);
```

Parameters:

pipe—pointer to pipe control block.

Returns:

NU_SUCCESS—indicates successful completion of the service.

NU_INVALID_PIPE—indicates the pipe pointer is invalid.

Pipe pointers

This API call builds a sequential list of pointers to all pipes in the system. It is not needed with Nucleus SE, as pipes are identified by a simple index, not a pointer, and it would be redundant. Service call prototype:

```
UNSIGNED NU_Pipe_Pointers(
NU_PIPE **pointer_list,
UNSIGNED maximum_pointers);
```

Parameters:

pointer_list—pointer to an array of NU_PIPE pointers; this array will be filled with pointers to established pipes in the system;

maximum_pointers—the maximum number of pointers to place in the array.

Returns:

The number of **NU_PIPE** pointers placed into the array

Broadcast to pipe

This API call broadcasts a message to all tasks waiting for a message from the specified pipe. It is not implemented with Nucleus SE, as it would have added excessive complexity.
Service call prototype:

```
STATUS NU_Broadcast_To_Pipe(
NU_PIPE *pipe,
VOID *message,
UNSIGNED size,
UNSIGNED suspend);
```

Parameters:

pipe—pointer to pipe control block;

message—pointer to the broadcast message;

size—the number of **UNSIGNED** data elements in the message. If the pipe supports variable-length messages, this parameter must be equal to or less than the message size supported by the pipe. If the pipe supports fixed-size messages, this parameter must be exactly the same as the message size supported by the pipe;

suspend—specifies whether or not to suspend the calling task if the pipe is already full; valid options for this parameter are **NU_NO_SUSPEND**, **NU_SUSPEND**, or a timeout value.

Returns:

NU_SUCCESS—indicates successful completion of the service.

NU_INVALID_PIPE—indicates the pipe pointer is invalid.

NU_INVALID_POINTER—indicates that the message pointer is **NULL**.

NU_INVALID_SIZE—indicates that the message size specified is not compatible with the size specified when the pipe was created.

NU_INVALID_SUSPEND—indicates that suspend attempted from a nontask thread.

NU_PIPE_FULL—indicates that there is insufficient space in the pipe for the message.

NU_TIMEOUT—indicates the pipe is still full after the timeout has expired.

NU_PIPE_DELETED—pipe was deleted, while the task was suspended.

NU_PIPE_RESET—pipe was reset, while the task was suspended.

Compatibility with Nucleus RTOS

With all aspects of Nucleus SE, it was my goal to maintain as high a level of applications code compatibility with Nucleus RTOS as possible. Pipes are no exception, and, from a user's perspective, they are implemented in much the same way as in Nucleus RTOS. There are areas of incompatibility, which have come about where I determined that such an incompatibility would be acceptable, given that the resulting code is easier to understand, or, more likely, could be made more memory efficient. Otherwise, Nucleus RTOS API calls may be almost directly mapped onto Nucleus SE calls. Chapter 20, *Using Nucleus SE*, includes further information on using Nucleus SE for users of Nucleus RTOS.

Object identifiers

In Nucleus RTOS, all objects are described by a data structure—a control block—which has a specific data type. A pointer to this control block serves as an identifier for the pipe. In Nucleus SE, I decided that a different approach was needed for memory efficiency, and all kernel objects are described by a number of tables in RAM and/or ROM. The size of these tables is determined by the number of each object type that is configured. The identifier for a specific object is simply an index into those tables. So, I have defined NUSE_PIPE as being equivalent to U8; a variable—not a pointer—of this type then serves as the pipe identifier. This is a small incompatibility, which is easily handled if code is ported to or from Nucleus RTOS. Object identifiers are normally just stored and passed around and not operated on in any way.

Nucleus RTOS also supports naming of pipes. These names are only used for target-based debug facilities. I omitted them from Nucleus SE to save memory.

Message size and variability

In Nucleus RTOS, a pipe may be configured to handle messages that are comprised of an arbitrary number of bytes of data. Likewise, in Nucleus SE. Nucleus RTOS also supports pipes with variable size messages, where only the maximum size is specified at creation time. Variable size messages are not supported by Nucleus SE.

Pipe size

The number of messages in a pipe in Nucleus SE is limited to 256, as all the index variables and constants are type $u8$. Nucleus RTOS is not limited in this way.

Unimplemented API calls

Nucleus RTOS supports ten service calls to work with pipes. Of these, four are not implemented in Nucleus SE. Details of these and of the decision to omit them may be found in *"Unimplemented API calls"* section earlier in this chapter.

14

System time

The concept of time in the context of a real-time operating system (RTOS) was introduced in Chapter 3, *RTOS Services and Facilities*, along with an idea of the facilities associated with time that likely to be available with an RTOS.

Time in an RTOS

In a real-time system, it is self-evident that time is an important resource, so it is unsurprising that a number of facilities of an RTOS are dedicated to different aspects of time.

Clock tick

All timing facilities are driven by a hardware clock. This is simply an oscillator that generates an interrupt at regular intervals. For timing numbers to be meaningful to applications programs, the frequency of the oscillator must be known.

Clock interrupt service routine

The interrupts generated by the hardware clock must be handled appropriately by an interrupt service routine (ISR), which implements all the timing facilities of an RTOS. The details of the clock ISR in Nucleus SE will be covered in Chapter 16, *Interrupts in Nucleus SE*.

Timing facilities

Nucleus RTOS and Nucleus SE include a number of timing facilities:

A tick clock—This is a simple counter that is incremented by clock ISR. In both Nucleus RTOS and Nucleus SE this counter is 32 bits wide and there are facilities for tasks to set and read its value. In Nucleus SE the tick clock is optional. This topic is discussed in more detail in the remainder of this chapter.

Embedded RTOS Design. DOI: https://doi.org/10.1016/B978-0-12-822851-7.00014-X

Application timers—Both Nucleus RTOS and Nucleus SE support timer objects. Their use and implementation in Nucleus SE are discussed in detail in Chapter 15, *Application Timers.*

Time slice scheduling—In Nucleus RTOS, tasks of the same priority are scheduled on a round-robin basis, but a time slice may also be applied. In Nucleus SE, a time slice scheduler is an option; this was discussed in detail in Chapter 5, *The Scheduler.*

Task sleep—A task may elect to suspend itself (go to sleep) for a specified time period. This facility was described in detail in Chapter 6, *Tasks.*

Application program interface (API) call timeouts—In both Nucleus RTOS and Nucleus SE a number of API calls allow a task to suspend, pending the availability of a resource. This suspension may be indefinite, or, in Nucleus RTOS, an optional timeout period may be specified. API call timeouts are not supported in Nucleus SE.

Accuracy

A brief word about timing accuracy is worthwhile at this point.

The precision of a timing facility is totally dependent on the frequency of the clock oscillator. For example, if a pulse is received every 10 ms and an application task wants to delay for 100 ms, it clearly needs to wait for 10 pulses. However, it is not known when the previous pulse occurred—it may have been just now, or it could have been nearly 10 ms ago. Hence a 100-ms delay could be as long as (just under) 110 ms.

An obvious way to solve this problem is to increase the oscillator frequency. If the pulse were to occur at 1-ms intervals, the 100-ms delay would never be longer than 101 ms. The downside is that 10 times as much CPU time would be spent in the clock ISR, which is an overhead.

The system designer must determine the balance between required timer accuracy and available CPU power.

Configuring system time

As with most aspects of Nucleus SE, the configuration of system time is primarily controlled by `#define` statements in `nuse_config.h`. The key setting is `NUSE_SYSTEM_TIME_SUPPORT`, which enables the facility. There is no question of specifying the number of objects—system time is simply enabled or not.

Choosing a nonzero value is the "master enable" for system time. This causes a data structure to be defined, of which more later in this chapter. It also activates the API enabling settings.

API enables

Every API function (service call) in Nucleus SE has an enabling `#define` symbol in `nuse_config.h`. For system time, these are:

```
NUSE_CLOCK_SET
NUSE_CLOCK_RETRIEVE
```

By default, both of these are set to `FALSE`, thus disabling each service call and inhibiting the inclusion of any implementation code. To configure system time for an application, you need to select the API calls that you want to use and set their enabling symbols to `TRUE`.

Here is an extract from the default `nuse_config.h` file.

```
#define NUSE_SYSTEM_TIME_SUPPORT FALSE     /* Enables the system
                           tick clock */

#define NUSE_CLOCK_SET FALSE     /* Service call enabler */
#define NUSE_CLOCK_RETRIEVE FALSE     /* Service call enabler */
```

A compile time error will result if a system time API function is enabled and the system time facility has not been enabled. If your code uses an API call, which has not been enabled, a link time error will result, as no implementation code will have been included in the application.

System time service calls

Nucleus RTOS supports two service calls that appertain to system time, which provide the following functionality:

Set the system time value. Implemented by `NUSE_Clock_Set()` in Nucleus SE.

Obtain the system time value. Implemented by `NUSE_Clock_Retrieve()` in Nucleus SE.

The implementation of each of these service calls is examined in detail.

System time set and obtain services

The only operations, which can be performed on the system time value, are to set it to a value and obtain (retrieve) its current value. Nucleus RTOS and Nucleus SE each provide two basic API calls for these operations, which will be discussed here.

The interpretation of the system time value is application-dependent, as it is simply a count of how many clock "ticks" there have been since the counter was last reset. To make practical use of this information, the frequency of the clock oscillator must be known.

Setting time

Any task may set system time by making a call to this API function.

Nucleus RTOS API call for setting time

Service call prototype:

```
VOID NU_Set_Clock(UNSIGNED new_value);
```

Parameters:

new_value—the value to which the system time is to be set

Returns:

Nothing

Nucleus SE API call for setting time

This API call supports the key functionality of the Nucleus RTOS API.

Service call prototype:

```
void NUSE_Clock_Set(U32 new_value);
```

Parameters:

new_value—the value to which the system time is to be set

Returns:

Nothing

Nucleus SE implementation of setting time

The code is very simple. The supplied value is just stored into `NUSE_Tick_Clock` within a critical section.

Retrieving time

A task may obtain system time by making a call to this API function.

Nucleus RTOS API call for retrieving time

Service call prototype:

```
UNSIGNED NU_Retrieve_Clock(VOID);
```

Parameters:

>   None

Returns:

>   The current system time value

Nucleus SE API call for retrieving time

This API call supports the key functionality of the Nucleus RTOS API.

Service call prototype:

```
U32 NUSE_Clock_Retrieve(void);
```

Parameters:

>   None

Returns:

>   The current system time value

Nucleus SE implementation of retrieving time

The code is very simple. The value of `NUSE_Tick_Clock` is obtained within a critical section and returned.

Data structures

System time utilizes one data structure—in RAM—which is just a single 32-bit word.

I strongly recommend that application code does not access this data structure directly but uses the provided API functions.

This avoids incompatibility with future versions of Nucleus SE and unwanted side effects and simplifies porting of an application to Nucleus RTOS. The details of data structures are included here to facilitate easier understanding of the working of the service call code and for debugging.

RAM data

The data structure is:

NUSE_Tick_Clock—This is a variable of type **u32** where the system clock tick count is stored.

This data structure is initialized to zero by **NUSE_Init_Task()** when Nucleus SE starts up. Chapter 17, *Nucleus SE Initialization and Start-up*, provides a full description of Nucleus SE start-up procedures.

ROM data

There are no ROM data structures associated with system time.

System time data footprint

Like all aspects of Nucleus SE, the amount of data memory required for system time is readily predictable.

The ROM data footprint is 0.

The RAM data footprint (in bytes) is always 4.

Unimplemented API calls

All Nucleus RTOS API calls that appertain to system time have an equivalent in Nucleus SE.

Compatibility with Nucleus RTOS

With all aspects of Nucleus SE, it was my goal to maintain as high a level of applications code compatibility with Nucleus RTOS as possible. System time is no exception, and, from a user's perspective, it is implemented in much the same way as in Nucleus RTOS. The Nucleus RTOS API calls may be directly mapped onto Nucleus SE calls.

15

Application timers

The idea of application timers was introduced in Chapter 3, *RTOS Services and Facilities*. They are kernel objects that provide tasks with simple means to time events or, more commonly, perform an activity on a regular basis. All the detail of timing functionality—accuracy, interrupt handling, etc.—in Nucleus SE was covered in Chapter 14, *System Time*.

Using timers

Application timers may be configured to be one-shot—that is, they are started and then, after the assigned time period, they simply terminate. Or a timer may be configured to repeat—that is, when it expires, it automatically restarts; the time interval for the restart period may be different from the initial time. A timer may also be optionally configured to run a specific function—an expiration routine—when (or each time) the timer expires.

Configuring timers

Number of timers

As with most aspects of Nucleus SE, the configuration of timers is primarily controlled by `#define` statements in `nuse_config.h`. The key setting is `NUSE_TIMER_NUMBER`, which determines how many timers are configured for the application. The default setting is 0 (i.e., no timers are in use) and you can set it to any value up to 16. An erroneous value will result in a compile time error, which is generated by a test in `nuse_config_check.h` (this is included into `nuse_config.c` and hence compiled with this module), resulting in a `#error` statement being compiled.

Choosing a nonzero value is the "master enable" for timers. This results in some data structures being defined and

Embedded RTOS Design. DOI: https://doi.org/10.1016/B978-0-12-822851-7.00015-1

sized accordingly, of which more later in this chapter. It also activates the application program interface (API) enabling settings.

Expiration routine enable

In Nucleus SE, I looked for opportunities to make functionality optional, where its omission would save memory. A good example of that is support for timer expiration routines. Apart from being optional for each individual timer, the facility may be enabled (or not) for the whole application by means of the `NUSE_TIMER_EXPIRATION_ROUTINE_SUPPORT` setting in `nuse_config.h`. Setting this to `FALSE` suppresses the definition of two ROM data structures, of which more later in this chapter.

API enables

Every API function (service call) in Nucleus SE has an enabling `#define` symbol in `nuse_config.h`. For timers, these are:

```
NUSE_TIMER_CONTROL
NUSE_TIMER_GET_REMAINING
NUSE_TIMER_RESET
NUSE_TIMER_INFORMATION
NUSE_TIMER_COUNT
```

By default, all of these are set to `FALSE`, thus disabling each service call and inhibiting the inclusion of any implementation code. To configure timers for an application, you need to select the API calls that you want to use and set their enabling symbols to `TRUE`.

Here is an extract from the default `nuse_config.h` file.

```
#define NUSE_TIMER_NUMBER   0   /* Number of application timers
                                   in the system - 0-16 */

          /* Service call enablers */
#define NUSE_TIMER_CONTROL      FALSE
#define NUSE_TIMER_GET_REMAINING      FALSE
#define NUSE_TIMER_RESET      FALSE
#define NUSE_TIMER_INFORMATION      FALSE
#define NUSE_TIMER_COUNT      FALSE
```

A compile time error will result if a timer API function is enabled and no timers are configured (except for

`NUSE_Timer_Count()`, which is always permitted). If your code uses an API call, which has not been enabled, a link time error will result, as no implementation code will have been included in the application.

Timer service calls

Nucleus RTOS supports eight service calls that appertain to timers, which provide the following functionality:

Control (start/stop) a timer. Implemented by `NUSE_Timer_Control()` in Nucleus SE.

Obtain remaining time from a timer. Implemented by `NUSE_Timer_Get_Remaining()` in Nucleus SE.

Restore a timer to the unused state (reset). Implemented by `NUSE_Timer_Reset()` in Nucleus SE.

Provide information about a specified timer. Implemented by `NUSE_Timer_Information()` in Nucleus SE.

Return a count of how many timers are (currently) configured for the application. Implemented by `NUSE_Timer_Count()` in Nucleus SE.

Add a new timer to the application (create). Not implemented in Nucleus SE.

Remove a timer from the application (delete). Not implemented in Nucleus SE.

Return pointers to all the timers (currently) in the application. Not implemented in Nucleus SE.

The implementation of each of these service calls is examined in detail.

Timer services

The fundamental operations, which can be performed on a timer, are controlling it—starting and stopping—and reading its current value. Nucleus RTOS and Nucleus SE each provide two basic API calls for these operations, which will be discussed here.

Controlling a timer

The Nucleus RTOS API call for controlling a timer simply permits the timer to be enabled or disabled (i.e., started or stopped). Nucleus SE provides the same service.

Nucleus RTOS API call for controlling a timer

Service call prototype:

```
STATUS NU_Control_Timer(
NU_TIMER *timer,
OPTION enable);
```

Parameters:

timer—pointer to the user-supplied timer control block;

enable—required function; may be NU_ENABLE_TIMER or NU_DISABLE_TIMER.

Returns:

NU_SUCCESS—The call was completed successfully.

NU_INVALID_TIMER—The timer pointer is invalid.

NU_INVALID_ENABLE—The specified function is invalid.

Nucleus SE API call for controlling a timer

This API call supports the full functionality of the Nucleus RTOS API.

Service call prototype:

```
STATUS NUSE_Timer_Control(
NUSE_TIMER timer,
OPTION enable);
```

Parameters:

timer—the index (ID) of the timer to be utilized;

enable—required function; may be NUSE_ENABLE_TIMER or NUSE_DISABLE_TIMER.

Returns:

NUSE_SUCCESS—The call was completed successfully.

NUSE_INVALID_TIMER—The timer index is invalid.

NUSE_INVALID_ENABLE—The specified function is invalid.

Nucleus SE implementation of timer control

The code of the NUSE_Timer_Control() API function—after parameter checking—is reasonably straightforward:

```
NUSE_CS_Enter();

if (enable == NUSE_ENABLE_TIMER)
{
  NUSE_Timer_Status[timer] = TRUE;
  if (NUSE_Timer_Expirations_Counter[timer] == 0)
  {
    NUSE_Timer_Value[timer] = NUSE_Timer_Initial_Time[timer];
  }
  else
  {
    NUSE_Timer_Value[timer] =
    NUSE_Timer_Reschedule_Time[timer];
  }
}
else /* enable == NUSE_DISABLE_TIMER */
{
  NUSE_Timer_Status[timer] = FALSE;
}

NUSE_CS_Exit();
```

If the specified function is NUSE_DISABLE_TIMER, the timer's status (NUSE_Timer_Status[] entry) is set to FALSE, which results in it being ignored by the clock interrupt service routine (ISR).

If NUSE_ENABLE_TIMER is specified, the timer counter (NUSE_Timer_Value[]) is set to NUSE_Timer_Initial_Time[], if the timer has never expired since last reset. Otherwise, it is set to NUSE_Timer_Reschedule_Time[]. Then the timer's status (NUSE_Timer_Status[] entry) is set to TRUE, which results in it being processed by the clock ISR.

Reading a timer

The Nucleus RTOS API call for getting the remaining time from a timer does just that—returns the number of ticks before it will expire. Nucleus SE provides the same service.

Nucleus RTOS API call for getting remaining time

Service call prototype:

```
STATUS NU_Get_Remaining_Time(
NU_TIMER *timer,
UNSIGNED *remaining_time);
```

Parameters:

> `timer`—pointer to the user-supplied timer control block;
>
> `remaining_time`—a pointer to storage for the remaining time value, which is a single variable of type `UNSIGNED`.

Returns:

> `NU_SUCCESS`—The call was completed successfully.
>
> `NU_INVALID_TIMER`—The timer pointer is invalid.

Nucleus SE API call for getting remaining time

This API call supports the full functionality of the Nucleus RTOS API.

Service call prototype:

```
STATUS NUSE_Timer_Get_Remaining(
NUSE_TIMER timer,
U16 *remaining_time);
```

Parameters:

> `timer`—the index (ID) of the timer to be utilized;
>
> `remaining_time`—a pointer to storage for the remaining time value, which is a single variable of type `U16`.

Returns:

> `NUSE_SUCCESS`—The call was completed successfully.
>
> `NUSE_INVALID_TIMER`—The timer index is invalid.
>
> `NUSE_INVALID_POINTER`—The remaining time pointer is `NULL`.

Nucleus SE implementation of timer read

The bulk of the code of the `NUSE_ Timer_Get_Remaining()` API function—after parameter checking—is almost trivially simple.

The value of NUSE_Timer_Value[] is obtained and returned within a critical section.

Timer utility services

Nucleus RTOS has four API calls that provide utility functions associated with timers: reset timer, return information about a timer, return number of timers in the application, and return pointers to all timers in the application. The first three of these are implemented in Nucleus SE.

Resetting a timer

This API call restores the timer to its initial, unused state. The timer may be enabled or disabled after the completion of this call. It is only allowed after a timer has been disabled (using NUSE_Timer_Control()). On the next occasion that the timer is enabled, it will be initialized with its entry from NUSE_Timer_Initial_Time[]. Nucleus RTOS permits new initial and reschedule times to be provided and the expiration routine to be specified when a timer is reset; in Nucleus SE, these values are set at configuration time and cannot be changed, as they may be stored in ROM.

Nucleus RTOS API call for resetting a timer

Service call prototype:

```
STATUS NU_Reset_Timer(
NU_TIMER *timer,
VOID (*expiration_routine)(UNSIGNED),
UNSIGNED initial_time,
UNSIGNED reschedule_time,
OPTION enable);
```

Parameters:

timer—pointer to the timer to be reset;

expiration_routine—specifies the application routine to execute when the timer expires;

initial_time—the initial number of timer ticks for timer expiration;

reschedule_time—the number of timer ticks for expiration after the first expiration;

`enable`—required state after reset; may be `NU_ENABLE_TIMER` or `NU_DISABLE_TIMER`.

Returns:

`NU_SUCCESS`—The call was completed successfully.

`NU_INVALID_TIMER`—The timer control block pointer is not valid.

`NU_INVALID_FUNCTION`—The expiration function pointer is `NULL`.

`NU_INVALID_ENABLE`—The specified state is invalid.

`NU_NOT_DISABLED`—The time is currently enabled (disable required before reset).

Nucleus SE API call for resetting a timer

This API call supports a simplified version of the key functionality of the Nucleus RTOS API.

Service call prototype:

```
STATUS NUSE_Timer_Reset(
NUSE_TIMER timer,
OPTION enable);
```

Parameters:

`timer`—the index (ID) of the timer to be reset;

`enable`—required state after reset; may be `NUSE_ENABLE_TIMER` or `NUSE_DISABLE_TIMER`.

Returns:

`NUSE_SUCCESS`—The call was completed successfully.

`NUSE_INVALID_TIMER`—The timer index is not valid.

`NUSE_INVALID_ENABLE`—The specified state is invalid.

`NUSE_NOT_DISABLED`—The time is currently enabled (disable required before reset).

Nucleus SE implementation of timer reset

The bulk of the code of the `NUSE_Timer_Reset()` API function—after parameter and current status checking—is quite straightforward:

```
NUSE_CS_Enter();

NUSE_Init_Timer(timer);

if (enable == NUSE_ENABLE_TIMER)
{
  NUSE_Timer_Status[timer] = TRUE;
}

/* else enable == NUSE_DISABLE_TIMER and status remains FALSE */

NUSE_CS_Exit();
```

`NUSE_Init_Timer()` is called, which initializes the time value and clears the expiration counter. A check is then made on required state and the timer enabled, if required.

Timer information

This service call obtains a selection of information about a timer. The Nucleus SE implementation differs from Nucleus RTOS in that it returns less information, as object naming is not supported.

Nucleus RTOS API call for timer information

Service call prototype:

```
STATUS NU_Timer_Information(
NU_TIMER *timer,
CHAR *name,
OPTION *enable,
UNSIGNED *expirations,
UNSIGNED *id,
UNSIGNED *initial_time,
UNSIGNED *reschedule_time);
```

Parameters:

timer—pointer to the timer about which information is being requested;

name—pointer to an 8-character destination area for the timer's name;

enable—a pointer to a variable, which will receive the timer's current enable state: `NU_ENABLE_TIMER` or `NU_DISABLE_TIMER`;

expirations—a pointer to a variable of type, which will receive the count of the number of times the timer has expired since it was last reset;

> `id`—a pointer to a variable, which will receive the value of the parameter passed to the timer's expiration routine;
>
> `initial_time`—a pointer to a variable, which will receive the value to which the timer is initialized on reset;
>
> `reschedule_time`—a pointer to a variable, which will receive the value to which the timer is initialized on expiration.

Returns:

> `NU_SUCCESS`—The call was completed successfully.
>
> `NU_INVALID_TIMER`—The timer pointer is not valid.

Nucleus SE API call for timer information

This API call supports the key functionality of the Nucleus RTOS API.

Service call prototype:

```
STATUS NUSE_Timer_Information(
NUSE_TIMER timer,
OPTION *enable,
U8 *expirations,
U8 *id,
U16 *initial_time,
U16 *reschedule_time);
```

Parameters:

> `timer`—the index of the timer about which information is being requested;
>
> `enable`—a pointer to a variable, which will receive a value of `NUSE_ENABLE_TIMER` or `NUSE_DISABLE_TIMER` depending on whether the timer is enabled or not;
>
> `expirations`—a pointer to a variable of type `U8`, which will receive the count of the number of times the timer has expired since it was last reset;
>
> `id`—a pointer to a variable of type `U8`, which will receive the value of the parameter passed to the timer's expiration routine (nothing returned if expiration routines are disabled);
>
> `initial_time`—a pointer to a variable of type `U16`, which will receive the value to which the timer is initialized on reset;

reschedule_time—a pointer to a variable of type u16, which will receive the value to which the timer is initialized on expiration.

Returns:

NUSE_SUCCESS—The call was completed successfully.

NUSE_INVALID_TIMER—The timer index is not valid.

NUSE_INVALID_POINTER—One or more of the pointer parameters is invalid.

Nucleus SE implementation of timer information

The implementation of this API call is quite straightforward:

```
NUSE_CS_Enter();

if (NUSE_Timer_Status[timer])
{
   *enable = NUSE_ENABLE_TIMER;
}
else
{
   *enable = NUSE_DISABLE_TIMER;
}

*expirations = NUSE_Timer_Expirations_Counter[timer];

#if NUSE_TIMER_EXPIRATION_ROUTINE_SUPPORT
   *id = NUSE_Timer_Expiration_Routine_Parameter[timer];
#endif

*initial_time = NUSE_Timer_Initial_Time[timer];
*reschedule_time = NUSE_Timer_Reschedule_Time[timer];

NUSE_CS_Exit();
```

The function returns the timer status. The value of the expiration routine parameter is only returned if expiration routines have been enabled in the application.

Obtaining the number of timers

This service call returns the number of timers configured in the application. Whilst in Nucleus RTOS this will vary over time and the returned value will represent the current number of timers, in Nucleus SE the value returned is set at build time and cannot change.

Nucleus RTOS API call for timer count

Service call prototype:

```
UNSIGNED NU_Established_Timers(VOID);
```

Parameters:

None

Returns:

The number of created timers in the system

Nucleus SE API call for timer count

This API call supports the key functionality of the Nucleus RTOS API.

Service call prototype:

```
U8 NUSE_Timer_Count(void);
```

Parameters:

None

Returns:

The number of configured timers in the application

Implementation of timer count

The implementation of this API call is almost trivially simple: the value of the `#define` symbol `NUSE_TIMER_NUMBER` is returned.

Data structures

Timers utilize five or seven data structures—in RAM and ROM—which, like other Nucleus SE objects, are a series of tables, included and dimensioned according to the number of timers configured and options selected.

I strongly recommend that application code does not access these data structures directly, but uses the provided API functions. This avoids incompatibility with future versions of Nucleus SE and unwanted side effects and simplifies porting of an application to Nucleus RTOS. The details of data structures are included here to facilitate easier understanding of the working of the service call code and for debugging.

RAM data

These data structures are:

NUSE_Timer_Status[]—This is an array of type U8, with an entry for each configured timer and is where the timer status (running or stopped: TRUE or FALSE) is stored.

NUSE_Timer_Value[]—This is an array of type U16, with one entry for each configured timer, which contains the current value of the timer's counter.

NUSE_Timer_Expirations_Counter[]—This type U8 array contains the counts of how many times the timers have reached expiration since they were last reset.

These data structures are all initialized by NUSE_Init_Timer() when Nucleus SE starts up. Chapter 17, *Nucleus SE Initialization and Start-up*, provides a full description of Nucleus SE start-up procedures.

Here are the definitions of these data structures in nuse_init.c file:

```
RAM U8 NUSE_Timer_Status[NUSE_TIMER_NUMBER];
RAM U16 NUSE_Timer_Value[NUSE_TIMER_NUMBER];
RAM U8 NUSE_Timer_Expirations_Counter[NUSE_TIMER_NUMBER];
```

ROM Data

These data structures are:

NUSE_Timer_Initial_Time[]—This is an array of type U16, with one entry for each configured timer and is where the initial value for each timer's counter is stored.

NUSE_Timer_Reschedule_Time[]—This is an array of type U16, with one entry for each configured timer, which contains the value to which each timer's counter should be set on expiration. A value of zero indicates that the timer is a "one shot" and should not be restarted automatically.

NUSE_Timer_Expiration_Routine_Address[]—This type ADDR array contains the address of the timers' expiration routines.

This array only exists if timer expiry routine support has been enabled.

`NUSE_Timer_Expiration_Routine_Parameter[]`—This type `U8` array contains the value of a parameter that will be passed to the timers' expiration routines. This array only exists if timer expiry routine support has been enabled.

These data structures are all declared and initialized (statically, of course) in `nuse_config.c`; thus:

```
ROM U16 NUSE_Timer_Initial_Time[NUSE_TIMER_NUMBER] =
{
  /* timer initial times ------ */
};

ROM U16 NUSE_Timer_Reschedule_Time[NUSE_TIMER_NUMBER] =
{
  /* timer reschedule times ------ */
};

#if NUSE_TIMER_EXPIRATION_ROUTINE_SUPPORT ||
                    NUSE_INCLUDE_EVERYTHING

  /* need prototypes of expiration routines here */

  ROM ADDR NUSE_Timer_Expiration_Routine_Address
                          [NUSE_TIMER_NUMBER] =
  {
    /* addresses of timer expiration routines ------ */
    /* can be NULL */
  };

  ROM U8
  NUSE_Timer_Expiration_Routine_Parameter[NUSE_TIMER_NUMBER] =
  {
    /* timer expiration routine parameters ------ */
  };

#endif
```

Timer data footprint

Like all kernel objects in Nucleus SE, the amount of data memory required for timers is readily predictable.

The RAM data footprint (in bytes) for all the timers in an application may be computed thus:

`NUSE_TIMER_NUMBER * 4`

The ROM data footprint (in bytes) for all the timers in an application, if expiration routines are not supported, may be computed thus:

```
NUSE_TIMER_NUMBER * 4
```

Otherwise, it is:

```
NUSE_TIMER_NUMBER * (sizeof(ADDR) + 5)
```

Unimplemented API calls

Three timer API calls found in Nucleus RTOS are not implemented in Nucleus SE.

Create timer

This API call creates a timer. It is not needed with Nucleus SE, as timers are created statically.

Service call prototype:

```
STATUS NU_Create_Timer(
NU_TIMER *timer,
CHAR *name,
VOID (*expiration_routine)(UNSIGNED),
UNSIGNED id,
UNSIGNED initial_time,
UNSIGNED reschedule_time,
OPTION enable);
```

Parameters:

timer—pointer to a user-supplied timer control block; this will be used as a "handle" for the timer in other API calls;

name—pointers to a 7-character, NULL-terminated name for the timer;

expiration_routine—specifies the application routine to execute when the timer expires;

id—an UNSIGNED data element supplied to the expiration routine; the parameter may be used to help identify timers that use the same expiration routine;

initial_time—specifies the initial number of timer ticks for timer expiration;

reschedule_time—specifies the number of timer ticks for expiration after the first expiration; if this parameter is zero, the timer only expires once;

enable—valid options for this parameter are NU_ENABLE_TIMER and NU_DISABLE_TIMER; NU_ENABLE_TIMER activates the timer after it is created; NU_DISABLE_TIMER leaves the timer disabled; timers created with the NU_DISABLE_TIMER must be enabled by a call to NU_Control_Timer() later.

Returns:

NU_SUCCESS—indicates successful completion of the service.

NU_INVALID_TIMER—indicates the timer control block pointer is NULL or already in use.

NU_INVALID_FUNCTION—indicates the expiration function pointer is NULL.

NU_INVALID_ENABLE—indicates that the enable parameter is invalid.

NU_INVALID_OPERATION—indicates that initial_time parameter was zero.

Delete timer

This API call deletes a previously created timer, which must be disabled. It is not needed with Nucleus SE, as timers are created statically and cannot be deleted.

Service call prototype:

```
STATUS NU_Delete_Timer(NU_TIMER *timer);
```

Parameters:

timer—pointer to timer control block.

Returns:

NU_SUCCESS—indicates successful completion of the service.

NU_INVALID_TIMER—indicates the timer pointer is invalid.

NU_NOT_DISABLED—indicates the specified timer is not disabled.

Timer pointers

This API call builds a sequential list of pointers to all timers in the system. It is not needed with Nucleus SE, as timers are identified by a simple index, not a pointer, and it would be redundant.

Service call prototype:

```
UNSIGNED NU_Timer_Pointers(
NU_TIMER **pointer_list,
UNSIGNED maximum_pointers);
```

Parameters:

pointer_list—pointer to an array of NU_TIMER pointers; this array will be filled with pointers to established timers in the system;

maximum_pointers—the maximum number of pointers to place in the array.

Returns:

The number of NU_TIMER pointers placed into the array

Compatibility with Nucleus RTOS

With all aspects of Nucleus SE, it was my goal to maintain as high a level of applications code compatibility with Nucleus RTOS as possible. Timers are no exception, and, from a user's perspective, they are implemented in much the same way as in Nucleus RTOS. There are areas of incompatibility, which have come about where I determined that such an incompatibility would be acceptable, given that the resulting code is easier to understand, or, more likely, could be made more memory efficient. Otherwise, Nucleus RTOS API calls may be almost directly mapped onto Nucleus SE calls. Chapter 20, *Using Nucleus SE*, includes further information on using Nucleus SE for users of Nucleus RTOS.

Object identifiers

In Nucleus RTOS, all objects are described by a data structure—a control block—which has a specific data type. A pointer to this control block serves as an identifier for the timer. In Nucleus SE, I decided that a different approach was needed for memory efficiency, and all kernel objects are described by a number of tables in RAM and/or ROM. The size of these tables is determined by the number of each object type that is configured. The identifier for a specific object is simply an index into those tables. So, I have defined NUSE_TIMER as being equivalent to u8; a variable—not a pointer—of this type then serves as the timer identifier. This is a small incompatibility, which is easily handled if code is ported to or from Nucleus RTOS. Object identifiers are normally just stored and passed around and not operated on in any way.

Nucleus RTOS also supports naming of timers. These names are only used for target-based debug facilities. I omitted them from Nucleus SE to save memory.

Timer size

In Nucleus RTOS, timers are implemented with 32-bit counters. I decided to reduce this to 16 bits in Nucleus SE. This change imparts significant increases in memory and execution time efficiency. Nucleus SE could be modified quite readily, if longer times were an application requirement.

Expiration routines

Nucleus SE implements expiration routines in much the same way as in Nucleus RTOS, except that they may be disabled entirely (which saves some memory) and they are statically defined. It is not possible to change the expiration routine when a timer is reset.

Unimplemented API calls

Nucleus RTOS supports eight service calls to work with timers. Of these, three are not implemented in Nucleus SE. Details of these and of the decision to omit them may be found in "Unimplemented API calls" section earlier in this chapter.

16

Interrupts in Nucleus SE

All modern microprocessors and microcontrollers have interrupt facilities of some sort. This capability is essential to provide the responsiveness required for many applications. Of course, being responsive and predictable is also a key objective behind the use of a real-time operating system (RTOS), so the two topics do potentially conflict slightly. Using interrupts might compromise the real-time integrity of the OS. This subject, and the resolution of this conflict, was covered in Chapter 2, *Multitasking and Scheduling*. Here, we will look at the interrupt handling strategy employed with Nucleus SE.

In all cases, interrupts are not controlled by Nucleus SE—they are processed when they occur according to priority and vectoring in the usual way. Their execution time is simply "stolen" from that available to run the mainline application code and the scheduler. Clearly this implies that all interrupt service routines (ISRs) should be simple, short, and fast.

Native and managed interrupts

Nucleus SE does offer two ways to handle interrupts: "native" ISRs are nothing special and have somewhat limited opportunity to interact with the OS (at least they do when the priority scheduler is selected); "managed" ISRs have a much wider range of application program interface (API) calls that may be made.

By means of some entry/exit macros, an ISR used with a Nucleus SE application may be designated to be native or managed.

Native interrupts

Nucleus SE native interrupts are standard ISRs—you can think of them as "unmanaged." They would typically be used when an interrupt may occur with high frequency and requires servicing with very low overhead. The routine is most likely to be coded in C, as many modern embedded compilers support the writing of ISRs

Embedded RTOS Design. DOI: https://doi.org/10.1016/B978-0-12-822851-7.00016-3

by means of the `interrupt` keyword. The only context information saved is that which the compiler deems necessary. This leads to significant limitations in what operations a native interrupt routine can perform, as we shall see shortly.

To construct a Nucleus SE native ISR, simply write the ISR in the usual way, including a call to the macro `NUSE_NISR_Enter()` at the beginning and a call to `NUSE_NISR_Exit()` at the end. These macros are defined in `nuse_types.h` and simply set the global variable `NUSE_Task_State` to `NUSE_NISR_CONTEXT`.

Managed interrupts

If you need greater flexibility in what operations can be performed by an ISR, a Nucleus SE managed interrupt may be the solution. The key difference from a native interrupt is the context save. Instead of just allowing the compiler to stack a few registers, a managed interrupt saves the complete task context (in its context block) on entry. The current task's context is then loaded from its context block at the end. This accommodates the possibility that the current task may be changed by the operation of the ISR code; this is entirely possible when the priority scheduler is in use. A complete description of Nucleus SE context save and restore was included in Chapter 5, *The Scheduler*.

Clearly the complete context save represents a higher overhead than the stacking of a few registers performed by a native interrupt. This is the price for extra flexibility and the reason why there is a choice of approaches to handling interrupts.

A managed interrupt is constructed using the macro `NUSE_MANAGED_ISR()`, which is defined in `nuse_types.h`. This macro constructs a function that contains the following sequence:

task context save

set `NUSE_Task_State` to `NUSE_MISR_CONTEXT`

call user-supplied ISR code function

restore `NUSE_Task_State` to previous setting

task context restore

The macro takes two parameters: a name for the interrupt, which is used as the function name for the constructed routine, and the name of the function containing the user-supplied ISR logic.

The Nucleus SE real time clock ISR, which is described later in this chapter, serves as an example of a managed ISR.

API calls from interrupt service routines

The range of API functions that may be called from a native or managed ISR depends on which scheduler has been selected. Broadly, the use of the priority scheduler provides many opportunities for the scheduler to be invoked as a result of an API function call, which would be a problem in a native ISR.

API calls from a native ISR with priority scheduler

A limited range of API function calls are permitted from a native ISR with the priority scheduler. This limitation is the result of the flexibility of the Nucleus SE API—many calls can result in a task being made *Ready* and it would not be possible for the scheduler to be called by a native ISR (as the task context is not saved). There is greater flexibility if task blocking is not enabled.

The following API calls are always permitted:

```
NUSE_Task_Current()
NUSE_Task_Check_Stack()
NUSE_Task_Information()
NUSE_Task_Count()
NUSE_Partition_Pool_Information()
NUSE_Partition_Pool_Count()
NUSE_Mailbox_Information()
NUSE_Mailbox_Count()
NUSE_Queue_Information()
NUSE_Queue_Count()
NUSE_Pipe_Information()
NUSE_Pipe_Count()
NUSE_Semaphore_Information()
NUSE_Semaphore_Count()
NUSE_Event_Group_Information()
NUSE_Event_Group_Count()
NUSE_Signals_Send()
NUSE_Timer_Control()
NUSE_Timer_Get_Remaining()
NUSE_Timer_Reset()
NUSE_Timer_Information()
NUSE_Timer_Count()
NUSE_Clock_Set()
NUSE_Clock_Retrieve()
NUSE_Release_Information()
```

However, the only one that is really useful is NUSE_Signals_Send(), as this provides a good way to indicate to a task that some work is required.

If blocking is disabled, which means that a task may not be made *Ready* by many API calls, a number of additional API functions are available:

```
NUSE_Partition_Allocate()
NUSE_Partition_Deallocate()
NUSE_Mailbox_Send()
NUSE_Mailbox_Receive()
NUSE_Mailbox_Reset()
NUSE_Queue_Send()
NUSE_Queue_Receive()
NUSE_Queue_Jam()
NUSE_Queue_Reset()
NUSE_Pipe_Send()
NUSE_Pipe_Receive()
NUSE_Pipe_Jam()
NUSE_Pipe_Reset()
NUSE_Semaphore_Obtain()
NUSE_Semaphore_Release()
NUSE_Semaphore_Reset()
NUSE_Event_Group_Set()
NUSE_Event_Group_Retrieve()
```

Certain API calls are never allowed from native ISRs, as they inevitably require the scheduler to operate:

```
NUSE_Task_Suspend()
NUSE_Task_Resume()
NUSE_Task_Sleep()
NUSE_Task_Relinquish()
NUSE_Task_Reset()
NUSE_Signals_Receive()
```

API calls from a managed ISR or native ISR with nonpriority scheduler

A much wider range of API functions may be called from an ISR when the run to completion, round robin, or time slice scheduler is in use. If the priority scheduler is used, a managed ISR facilitates a similar wide range. This is because calls that may result in a different task being scheduled are permitted. This capability is facilitated by code in NUSE_Reschedule() that detects that the context of the call is an ISR and suppresses the context switch (allowing it to occur at the end of the ISR). Full details of the scheduler operation were covered in Chapter 5, *The Scheduler.*

A key requirement is that API calls within an ISR must not result in suspension of the current task—waiting on a resource, for example. In other words, such calls should be made with the suspend option set to NUSE_NO_SUSPEND.

Given this proviso, the following API calls may be used:

```
NUSE_Task_Current()
NUSE_Task_Check_Stack()
NUSE_Task_Information()
NUSE_Task_Count()
NUSE_Task_Suspend()
NUSE_Task_Resume()
NUSE_Task_Reset()
NUSE_Partition_Allocate()
NUSE_Partition_Deallocate()
NUSE_Partition_Pool_Information()
NUSE_Partition_Pool_Count()
NUSE_Mailbox_Send()
NUSE_Mailbox_Receive()
NUSE_Mailbox_Reset()
NUSE_Mailbox_Information()
NUSE_Mailbox_Count()
NUSE_Queue_Send()
NUSE_Queue_Receive()
NUSE_Queue_Jam()
NUSE_Queue_Reset()
NUSE_Queue_Information()
NUSE_Queue_Count()
NUSE_Pipe_Send()
NUSE_Pipe_Receive()
NUSE_Pipe_Jam()
NUSE_Pipe_Reset()
NUSE_Pipe_Information()
NUSE_Pipe_Count()
NUSE_Semaphore_Obtain()
NUSE_Semaphore_Release()
NUSE_Semaphore_Reset()
NUSE_Semaphore_Information()
NUSE_Semaphore_Count()
NUSE_Event_Group_Set()
NUSE_Event_Group_Retrieve()
NUSE_Event_Group_Information()
NUSE_Event_Group_Count()
NUSE_Signals_Send()
NUSE_Timer_Control()
NUSE_Timer_Get_Remaining()
NUSE_Timer_Reset()
NUSE_Timer_Information()
NUSE_Timer_Count()
NUSE_Clock_Set()
NUSE_Clock_Retrieve()
NUSE_Release_Information()
```

A few API calls are never permitted, as they appertain specifically to the current task:

```
NUSE_Task_Relinquish()
NUSE_Signals_Receive()
NUSE_Task_Sleep()
```

Real time clock ISR

The real time clock ISR is the only complete ISR provided with Nucleus SE. Apart from providing all the required timing functionality for Nucleus SE, it also serves as an example of how to code a managed interrupt.

Real time clock ISR operations

The facilities provided by the real time clock ISR were outlined in an earlier, which covered the broad topic of system time in Nucleus SE. All of the functionality is optional, depending on how the application is configured. Here is the complete code for the real time clock ISR.

```
#if NUSE_TIMER_NUMBER! = 0

{
  U8 timer;

  for (timer = 0; timer < NUSE_TIMER_NUMBER; timer++)
  {
    if (NUSE_Timer_Status[timer])
    {
      if (--NUSE_Timer_Value[timer] == 0)
      {
        NUSE_Timer_Expirations_Counter[timer]++;

        #if NUSE_TIMER_EXPIRATION_ROUTINE_SUPPORT ||
                                    NUSE_INCLUDE_EVERYTHING
          if (NUSE_Timer_Expiration_Routine_Address[timer]
            ! = NULL)
          {
          ((PF1)NUSE_Timer_Expiration_Routine_Address[timer])
          NUSE_Timer_Expiration_Routine_Parameter[timer]);
          }
        #endif
        /* reschedule? */
        if (NUSE_Timer_Reschedule_Time[timer]! = 0)
        {         /* yes: set up time */
          NUSE_Timer_Value[timer] =
          NUSE_Timer_Reschedule_Time[timer];
        }
```

```
           else
           {          /* no: disable */
             NUSE_Timer_Status[timer] = FALSE;
           }
         }
       }
     }
   }

#endif

#if NUSE_SYSTEM_TIME_SUPPORT || NUSE_INCLUDE_EVERYTHING
  NUSE_Tick_Clock++;

#endif

#if NUSE_TASK_SLEEP || NUSE_INCLUDE_EVERYTHING

{
  U8 task;

  for (task = 0; task < NUSE_TASK_NUMBER; task++)
  {
    if (NUSE_Task_Timeout_Counter[task] != 0)
    {
      NUSE_Task_Timeout_Counter[task]--;
      if (NUSE_Task_Timeout_Counter[task] == 0)
      {
        NUSE_Wake_Task(task);
      }
    }
  }
}

#endif

#if NUSE_SCHEDULER_TYPE == NUSE_TIME_SLICE_ SCHEDULER

  if (--NUSE_Time_Slice_Ticks == 0)
  {
    NUSE_Reschedule();
  }

#endif
```

We will look at each of the four areas of functionality in the real time clock ISR:

Timers

If any application timers are configured, the ISR loops around to service each one by decrementing its counter value.

If a timer expires (i.e., the counter reaches zero), two actions are effected:

1. If timer expiration routines are configured and the timer has a valid (not `NULL`) pointer to a function (in the `NUSE_Timer_Expiration_Routine_Address[]`), the routine is executed, receiving a parameter from `NUSE_Timer_Expiration_Routine_Parameter[]`.

2. If the timer has a reschedule time (i.e., a nonzero value in `NUSE_Timer_Reschedule_Time[]`), the timer is reloaded with that value.

Application timers were described in more detail in Chapter 15, *Application Timers*.

System clock

If a system clock is configured, the value of `NUSE_Tick_Clock` is simply incremented. Further discussion on system time may be found in Chapter 14, *System Time*.

Task sleep

If task sleep is enabled (i.e., the API call `NUSE_Task_Sleep()` is configured), each task's timeout counter (entry in `NUSE_Task_Timeout_Counter[]`) is checked and, if nonzero, decremented. If any counter reaches zero, the corresponding task is woken up.

Time slice scheduling

If the time slice scheduler is in use, the time slice counter (`NUSE_Time_Slice_Ticks`) is decremented. If it reaches zero, the scheduler is called. The call to `NUSE_Reschedule()` takes care of resetting the counter.

A managed interrupt

Some explanation of the reasons why the real time clock ISR is a managed interrupt may be useful, as, under the right circumstances, the user may want to recode it as a native interrupt to reduce the overhead. For example, if only the system time facility is used (i.e., no application timers, no task sleep, and not the time slice scheduler), a native interrupt would be fine. The needs for a managed interrupt are as follows:

If timers are used and expiration routines configured, these routines may make API calls (from the interrupt context) that will cause a reschedule. These are subject to the same limitations as API calls made from ISRs (see earlier in this chapter).

If the priority scheduler is in use, an expiration of task sleep may require the scheduling of a higher priority task.

If the time slice scheduler is in use, it will definitely be called from the real time clock ISR, so a managed interrupt is mandatory.

Nucleus RTOS compatibility

Since interrupts are implemented in a very different way in Nucleus SE, compared with Nucleus RTOS, no particular compatibility should be expected. Nucleus RTOS has a native/low-level/high-level interrupt scheme, which is somewhat analogous to native and managed interrupts in Nucleus SE.

Low- and high-level ISRs

Low-level ISR

A low-level interrupt service routine (LISR) executes as a normal ISR, which includes using the current stack. Nucleus RTOS saves context before calling an LISR and restores context after the LISR returns. Therefore LISRs may be written in C and may call other C routines. However, there are only a few Nucleus RTOS services available to an LISR. If the interrupt processing requires additional Nucleus RTOS services, a high-level interrupt service routine (HISR) must be activated. Nucleus RTOS supports nesting of multiple LISRs.

High-level ISR

HISRs are created and deleted dynamically. Each HISR has its own stack space and its own control block. The memory for each is supplied by the application. Of course, the HISR must be created before it is activated by an LISR.

Since an HISR has its own stack and control block, it can be temporarily blocked if it tries to access a Nucleus RTOS data structure that is already being accessed.

There are three priority levels available to HISRs. If a higher priority HISR is activated during processing of a lower priority HISR, the lower priority HISR is preempted in much the same manner as a task gets preempted. HISRs of the same priority are executed in the order in which they were originally activated. All activated HISRs are processed before normal task scheduling is resumed.

Nucleus RTOS interrupt API calls

Nucleus RTOS has a number of API calls to support its interrupt structure. None of these are implemented in Nucleus SE. For native interrupts, API calls provide the following facilities:

Control (enable/disable) an interrupt (locally and globally).

Set up interrupt vector.

For low-level interrupts:

Register an LISR with the kernel.

For high-level interrupts:

Create/delete high-level interrupts.

Activate a high-level interrupt.

Obtain the number of high-level interrupts (currently) in the application.

Obtain pointers to the control blocks for all the high-level interrupts.

Obtain the pointer to the control block of the current high-level interrupt.

Obtain information about a high-level interrupt.

Controlling an interrupt globally

This service enables or disables interrupts in a task-independent manner. Therefore an interrupt disabled by this service remains disabled until enabled by a subsequent call to this service.

Service call prototype:

```
INT NU_Control_Interrupts(INT new_level);
```

Parameters:

new_level—new interrupt level for the system. The options NU_DISABLE_INTERRUPTS (disable all interrupts) and NU_ENABLE_INTERRUPTS (enable all interrupts) are always available. Other options may be available depending on architecture.

Returns:

This service returns the previous level of enabled interrupts.

Controlling an interrupt locally

This service enables or disables interrupts in a task-dependent manner. This service changes the Status Register to the value specified. The Status Register will be set back to value set by the last call to `NU_Control_Interrupts()` on the next context switch.

Service call prototype:

```
INT NU_Local_Control_Interrupts(INT new_level);
```

Parameters:

`new_level`—new interrupt level for the current task. The options `NU_DISABLE_INTERRUPTS` (disable all interrupts) and `NU_ENABLE_INTERRUPTS` (enable all interrupts) are always available. Other options may be available depending on architecture.

Returns:

This service returns the previous level of enabled interrupts.

Set up an interrupt vector

This service replaces the interrupt vector specified by vector with a custom ISR.

Service call prototype:

```
VOID *NU_Setup_Vector(
INT vector,
VOID *new);
```

Parameters:

`vector`—the interrupt vector at which to register the interrupt;

`new`—the ISR to register at the vector.

Returns:

This service returns a pointer to the ISR previously registered at the interrupt vector.

Register an LISR interrupt

This service associates the LISR function with an interrupt vector. System context is automatically saved before calling the specified LISR and is restored after the LISR returns.

Service call prototype:

```
STATUS NU_Register_LISR(
INT vector,
VOID(*lisr_entry)(INT),
VOID (**old_lisr)(INT));
```

Parameters:

vector—the interrupt vector at which to register the interrupt;

lisr_entry—the function to register at the vector; a value of NU_NULL clears the vector;

old_lisr—the subroutine previously registered at the specified vector.

Returns:

NU_SUCCESS—successful completion of the service.

NU_INVALID_VECTOR—The specified vector is invalid.

NU_NOT_REGISTERED—The vector is not currently registered as de-registration was specified by lisr_entry.

NU_NO_MORE_LISRS—The maximum number of registered LISRs has been exceeded.

Create an HISR

This service creates an HISR.

Service call prototype:

```
STATUS NU_Create_HISR(
NU_HISR *hisr,
CHAR *name,
VOID (*hisr_entry)(VOID),
OPTION priority,
VOID *stack_pointer,
UNSIGNED stack_size);
```

Parameters:

> `hisr`—pointer to a user-supplied HISR control block;
>
> `name`—pointer to a 7-character, `NULL`-terminated name for the HISR;
>
> `hisr_entry`—the function entry point of the HISR;
>
> `priority`—there are three HISR priorities (0–2); priority 0 is the highest;
>
> `stack_pointer`—pointer to the HISR's stack area;
>
> `stack_size`—number of bytes in the HISR stack.

Returns:

> `NU_SUCCESS`—successful completion of the service.
>
> `NU_INVALID_HISR`—The HISR control block pointer is `NULL` or is already in use.
>
> `NU_INVALID_ENTRY`—The HISR entry pointer is `NULL`.
>
> `NU_INVALID_PRIORITY`—The HISR priority is invalid.
>
> `NU_INVALID_MEMORY`—The stack pointer is `NULL`.
>
> `NU_INVALID_SIZE`—The stack size is too small.

Delete an HISR

This service deletes a previously created HISR.

Service call prototype:

```
STATUS NU_Delete_HISR(NU_HISR *hisr);
```

Parameters:

> `hisr`—pointer to a user-supplied HISR control block.

Returns:

> `NU_SUCCESS`—successful completion of the service.
>
> `NU_INVALID_HISR`—The HISR pointer is invalid.

Activate an HISR

This service activates an HISR. If the specified HISR is currently executing, this activation request is not processed until the current execution is complete. An HISR is executed once for each activation request.

Service call prototype:

```
STATUS NU_Activate_HISR (NU_HISR *hisr);
```

Parameters:

hisr—pointer to the HISR control block.

Returns:

NU_SUCCESS—successful completion of the service.

NU_INVALID_HISR—The HISR control block pointer is not valid.

Obtain the number of HISRs in a system

This service returns the number of established HISRs. All created HISRs are considered established. Deleted HISRs are no longer considered established.

Service call prototype:

```
UNSIGNED NU_Established_HISRs(VOID);
```

Parameters:

None.

Returns:

This service call returns the number of established HISRs in the system.

Obtain pointers to HISR control blocks

This service builds a sequential list of pointers to all established HISRs in the system.

Service call prototype:

```
UNSIGNED NU_HISR_Pointers(
NU_HISR **pointer_list,
UNSIGNED maximum_pointers);
```

Parameters:

> `pointer_list`—pointer to an array of `NU_HISR` pointers; this array will be filled with pointers of established HISRs in the system;
>
> `maximum_pointers`—the maximum number of `NU_HISR` pointers to place into the array; typically, this will be the size of the `pointer_list` array.

Returns:

> This service call returns the number of HISRS that are active in the system.

Obtain a pointer to the current HISR

This service returns the currently executing HISR's pointer.

Service call prototype:

```
NU_HISR *NU_Current_HISR_Pointer(VOID);
```

Parameters:

> None.

Returns:

> This service call returns a pointer the currently executing HISR's control block. If the caller is not an HISR, the value returned is `NU_NULL`.

Obtain information about an HISR

This service returns various information about the specified HISR.

Service call prototype:

```
STATUS NU_HISR_Information(
NU_HISR *hisr,
char *name,
UNSIGNED *scheduled_count,
DATA_ELEMENT *priority,
VOID **stack_base,
UNSIGNED *stack_size,
UNSIGNED *minimum_stack);
```

Parameters:

> `hisr`—pointer to the HISR;

`name`—pointer to an 8-character destination area for the HISR's name; this includes space for the `NULL` terminator;

`scheduled_count`— pointer to a variable for holding the total number of times this HISR has been scheduled;

`priority`—pointer to a variable for holding the HISR's priority;

`stack_base`—pointer to a pointer for holding the original stack pointer; this is the same pointer supplied during creation of the HISR;

`stack_size`—pointer to a variable for holding the total size of the HISR's stack;

`minimum_stack`—pointer to a variable for holding the minimum amount of available stack space detected during HISR execution.

Returns:

`NU_SUCCESS`—Successful completion of the service.

`NU_INVALID_HISR`—The HISR pointer is invalid.

API calls from ISRs

API calls from LISRs

An LISR may only make use of the following Nucleus RTOS services:

```
NU_Activate_HISR()
NU_Local_Control_Interrupts()
NU_Current_HISR_Pointer()
NU_Current_Task_Pointer()
NU_Retrieve_Clock()
```

API calls from HISRs

HISRs are allowed access to most Nucleus RTOS services, with the exception of self-suspension services. Additionally, since an HISR cannot suspend on a Nucleus RTOS service, the `suspend` parameter must always be set to `NU_NO_SUSPEND`.

17

Nucleus SE initialization and start-up

For any kind of operating system (OS), there is some type of start-up mechanism. Exactly how this works varies from one system to another. It is usual to say that an OS will "boot." This is an abbreviation for "bootstrap," which is a description of how a CPU gets from having a memory full of nothing in particular to a stable program execution state. Classically, a small piece of software is loaded into memory; it may simply be held in ROM. In years past, it may have been keyed in from the switches on the front panel of the computer. This "boot loader" would read in a more sophisticated bootstrap program, which, in turn, would load and start the OS. This is the process whereby a desktop computer gets started today; code in the BIOS seeks bootable devices (hard drives or CD-ROMs) from which a bootstrap and, hence, an OS is loaded.

An OS for an embedded system may also be initialized in this way. Indeed, embedded OSes, which are derived from desktop OSes, do exactly that. But for most "classic" real-time operating systems (RTOSes), a much simpler (and hence faster) process is used.

An OS is just a piece of software. If that software is already in memory—in some form of ROM, for example, it is simply a matter of arranging for the CPU's reset sequence to end up with the execution of the OS's initialization code. This is how most RTOSes work and Nucleus SE is no exception.

Most embedded software development toolkits include the necessary start-up code to handle a CPU reset and arrive at the entry point to the `main()` function. The Nucleus SE distribution code does not concern itself with this process, as it is intended to be as portable as possible. Instead, it provides a `main()` function, which takes control of the CPU and initializes and starts the OS; this is described in detail later in this chapter.

Memory initialization

The declarations of all the static variables in the Nucleus SE code are prefixed with ROM or RAM to indicate where they might

Embedded RTOS Design. DOI: https://doi.org/10.1016/B978-0-12-822851-7.00017-5

be sensibly located. These two #define symbols are defined in nuse_types.h and should be set up to accommodate the capabilities of the development toolkit (compiler and linker) in use. Typically, ROM may be set to const and RAM left blank.

All ROM variables are statically initialized, which is logical. No RAM variables are statically initialized (as this will only work with certain toolkits, which arrange for an automatic copy from ROM to RAM); explicit initialization code is included, of which more in the course of this chapter.

Nucleus SE does not keep any "constant" data in RAM, which, in small systems, may be in short supply. Instead of using complex data structures to describe kernel objects, a series of tables (arrays) are employed, which are easily located in ROM or RAM, as appropriate.

The main() function

Here is the complete code for the Nucleus SE main() function:

```
void main(void)
{
  NUSE_Init();    /* initialize kernel data */

  /* user initialization code here */

  NUSE_Scheduler();   /* start tasks */
}
```

The sequence of operations is quite straightforward:

The NUSE_Init() function is called first. This initializes all the Nucleus SE data structures and is outlined in more detail in the following section.

Next, there is the opportunity for the user in insert any application-specific initialization code, which will be executed prior to the start of the task scheduler. More details on what can be achieved by this code may be found later in this chapter.

Lastly, the Nucleus SE scheduler (NUSE_Scheduler()) is started. This is also examined in more detail later in this chapter.

The NUSE_Init() function

This function initializes all the Nucleus SE kernel variables and data structures. Here is the complete code:

```
void NUSE_Init(void)
{
  U8 index;

  /* global data */

  NUSE_Task_Active = 0;

  NUSE_Task_State = NUSE_STARTUP_CONTEXT;

  #if NUSE_SYSTEM_TIME_SUPPORT
    NUSE_Tick_Clock = 0;
  #endif

  #if NUSE_SCHEDULER_TYPE == NUSE_TIME_SLICE_SCHEDULER
    NUSE_Time_Slice_Ticks = NUSE_TIME_SLICE_TICKS;
  #endif

  /* tasks */

  #if ((NUSE_SCHEDULER_TYPE !=
  NUSE_RUN_TO_COMPLETION_SCHEDULER)
    || NUSE_SIGNAL_SUPPORT || NUSE_TASK_SLEEP
    || NUSE_SUSPEND_ENABLE || NUSE_SCHEDULE_COUNT_SUPPORT)
    for (index = 0; index < NUSE_TASK_NUMBER; index++)
    {
      NUSE_Init_Task(index);
    }
  #endif

  /* partition pools */

  #if NUSE_PARTITION_POOL_NUMBER! = 0
    for (index = 0; index < NUSE_PARTITION_POOL_NUMBER; index++)
    {
      NUSE_Init_Partition_Pool(index);
    }
  #endif

  /* mailboxes */

  #if NUSE_MAILBOX_NUMBER! = 0
    for (index = 0; index < NUSE_MAILBOX_NUMBER; index++)
    {
      NUSE_Init_Mailbox(index);
    }
  #endif

  /* queues */

  #if NUSE_QUEUE_NUMBER! = 0
    for (index = 0; index < NUSE_QUEUE_NUMBER; index++)
    {
      NUSE_Init_Queue(index);
    }
```

```
      #endif

      /* pipes */

      #if NUSE_PIPE_NUMBER! = 0
        for (index = 0; index < NUSE_PIPE_NUMBER; index++)
        {
          NUSE_Init_Pipe(index);
        }
      #endif

      /* semaphores */

      #if NUSE_SEMAPHORE_NUMBER! = 0
        for (index = 0; index < NUSE_SEMAPHORE_NUMBER; index++)
        {
          NUSE_Init_Semaphore(index);
        }
      #endif

      /* event groups */

      #if NUSE_EVENT_GROUP_NUMBER! = 0
        for (index = 0; index < NUSE_EVENT_GROUP_NUMBER; index++)
        {
          NUSE_Init_Event_Group(index);
        }
      #endif

      /* timers */

      #if NUSE_TIMER_NUMBER! = 0
        for (index = 0; index < NUSE_TIMER_NUMBER; index++)
        {
          NUSE_Init_Timer(index);
        }
      #endif
    }
```

First, some global variables are initialized:

NUSE_Task_Active—the index of the currently active task—is set to zero; this may be modified by the scheduler in due course.

NUSE_Task_State is set to NUSE_STARTUP_CONTEXT, which indicates the limited application program interface (API) functionality to any following application initialization code.

If system time support is enabled, NUSE_Tick_Clock is set to zero.

If the time slice scheduler has been enabled, NUSE_Time_Slice_Ticks is set up to the configured time slice value, NUSE_TIME_SLICE_TICKS.

Then, a sequence of functions is called to initialize kernel objects:

`NUSE_Init_Task()` is called to initialize data structures for each task. This call is only omitted if the run-to-completion scheduler is selected and signals, task suspend, and schedule counting are all not configured (as this combination would result in there being no RAM data structures appertaining to tasks and, hence, no initialization to be done).

`NUSE_Init_Partition_Pool()` is called to initialize each partition pool object. The calls are omitted if no partition pools have been configured.

`NUSE_Init_Mailbox()` is called to initialize each mailbox object. The calls are omitted if no mailboxes have been configured.

`NUSE_Init_Queue()` is called to initialize each queue object. The calls are omitted if no queues have been configured.

`NUSE_Init_Pipe()` is called to initialize each pipe object. The calls are omitted if no pipes have been configured.

`NUSE_Init_Semaphore()` is called to initialize each semaphore object. The calls are omitted if no semaphores have been configured.

`NUSE_Init_Event_Group()` is called to initialize each event group object. The calls are omitted if no event groups have been configured.

`NUSE_Init_Timer()` is called to initialize each timer object. The calls are omitted if no timers have been configured.

Initializing tasks

Here is the complete code for `NUSE_Init_Task()`:

```
void NUSE_Init_Task(NUSE_TASK task)

{
  #if NUSE_SCHEDULER_TYPE!=
  NUSE_RUN_TO_COMPLETION_SCHEDULER
    NUSE_Task_Context[task][15] =                 /* SR */
      NUSE_STATUS_REGISTER;
    NUSE_Task_Context[task][16] =                 /* PC */
      NUSE_Task_Start_Address[task];
    NUSE_Task_Context[task][17] =                 /* SP */
      (U32 *)NUSE_Task_Stack_Base[task] +
      NUSE_Task_Stack_Size[task];
  #endif
```

```
#if NUSE_SIGNAL_SUPPORT || NUSE_INCLUDE_EVERYTHING
  NUSE_Task_Signal_Flags[task] = 0;
#endif

#if NUSE_TASK_SLEEP || NUSE_INCLUDE_EVERYTHING
  NUSE_Task_Timeout_Counter[task] = 0;
#endif

#if NUSE_SUSPEND_ENABLE || NUSE_INCLUDE_EVERYTHING
  #if NUSE_INITIAL_TASK_STATE_SUPPORT ||
    NUSE_INCLUDE_EVERYTHING
    NUSE_Task_Status[task] =
      NUSE_Task_Initial_State[task];
  #else
    NUSE_Task_Status[task] = NUSE_READY;
  #endif
#endif

#if NUSE_SCHEDULE_COUNT_SUPPORT || NUSE_INCLUDE_EVERYTHING
  NUSE_Task_Schedule_Count[task] = 0;
#endif
}
```

Unless the run-to-completion scheduler has been configured, the context block—NUSE_Task_Context[task][]—for the task is initialized. Most entries are not set to a value, as they represent general machine registers that are assumed to have an indeterminate value when the task starts up. In the example (Freescale ColdFire) implementation of Nucleus SE (and this would be similar for any processor) the last three entries are set up explicitly:

NUSE_Task_Context[task][15] holds the status register (SR) and is set to the value in the #define symbol NUSE_STATUS_REGISTER.

NUSE_Task_Context[task][16] holds the program counter (PC) and is set to the address of the entry point of the task's code: NUSE_Task_Start_Address[task].

NUSE_Task_Context[task][17] holds the stack pointer (SP), which is initialized to a value computed by adding the address of the task's stack base (NUSE_Task_Stack_Base[task]) to the task's stack size (NUSE_Task_Stack_Size[task]).

If signal support is enabled, the task's signal flags (NUSE_Task_Signal_Flags[task]) are set to zero.

If task sleep (i.e., the API call NUSE_Task_Sleep()) is enabled, the task's timeout counter (NUSE_Task_Timeout_Counter[task]) is set to zero.

If task suspend is enabled, the task's status (NUSE_Task_Status[task]) is initialized. This initial value is user specified (in NUSE_Task_Initial_State[task]), if task initial task state support is enabled. Otherwise, the status is set to NUSE_READY.

If task schedule counting is enabled, the task's counter (NUSE_Task_Schedule_Count[task]) is set to zero.

Initializing partition pools

Here is the complete code for NUSE_Init_Partition_Pool():

```
void NUSE_Init_Partition_Pool(NUSE_PARTITION_POOL pool)
{
    NUSE_Partition_Pool_Partition_Used[pool] = 0;

    #if NUSE_BLOCKING_ENABLE

        NUSE_Partition_Pool_Blocking_Count[pool] = 0;

    #endif
}
```

The partition pool's "used" counter (NUSE_Partition_Pool_Partition_Used[pool]) is set to zero.

If task blocking is enabled, the partition pool's blocked task counter (NUSE_Partition_Pool_Blocking_Count[pool]) is set to zero.

Initializing mailboxes

Here is the complete code for NUSE_Init_Mailbox():

```
void NUSE_Init_Mailbox(NUSE_MAILBOX mailbox)
{
    NUSE_Mailbox_Data[mailbox] = 0;
    NUSE_Mailbox_Status[mailbox] = 0;

    #if NUSE_BLOCKING_ENABLE

        NUSE_Mailbox_Blocking_Count[mailbox] = 0;

    #endif
}
```

The mailbox's data store (NUSE_Mailbox_Data[mailbox]) is set to zero and its status (NUSE_Mailbox_Status[mailbox]) is set to "unused" (i.e., zero).

If task blocking is enabled, the mailbox's blocked task counter (NUSE_Mailbox_Blocking_Count[mailbox]) is set to zero.

Initializing queues

Here is the complete code for `NUSE_Init_Queue()`:

```
void NUSE_Init_Queue(NUSE_QUEUE queue)
{
  NUSE_Queue_Head[queue] = 0;
  NUSE_Queue_Tail[queue] = 0;
  NUSE_Queue_Items[queue] = 0;

  #if NUSE_BLOCKING_ENABLE
    NUSE_Queue_Blocking_Count[queue] = 0;
  #endif
}
```

The queue's head and tail pointers (actually, they are indexes—`NUSE_Queue_Head[queue]` and `NUSE_Queue_Tail[queue]`) are set to point to the start of the queue data area (i.e., given the value zero). The queue's item counter (`NUSE_Queue_Items[queue]`) is also set to zero.

If task blocking is enabled, the queue's blocked task counter (`NUSE_Queue_Blocking_Count[queue]`) is set to zero.

Initializing pipes

Here is the complete code for `NUSE_Init_Pipe()`:

```
void NUSE_Init_Pipe(NUSE_PIPE pipe)
{
  NUSE_Pipe_Head[pipe] = 0;
  NUSE_Pipe_Tail[pipe] = 0;
  NUSE_Pipe_Items[pipe] = 0;

  #if NUSE_BLOCKING_ENABLE

    NUSE_Pipe_Blocking_Count[pipe] = 0;

  #endif
}
```

The pipe's head and tail pointers (actually, they are indexes—`NUSE_Pipe_Head[pipe]` and `NUSE_Pipe_Tail[pipe]`) are set to point to the start of the pipe data area (i.e., given the value zero). The pipe's item counter (`NUSE_Pipe_Items[pipe]`) is also set to zero.

If task blocking is enabled, the pipe's blocked task counter (`NUSE_Pipe_Blocking_Count[pipe]`) is set to zero.

Initializing semaphores

Here is the complete code for `NUSE_Init_Semaphore()`:

```
void NUSE_Init_Semaphore(NUSE_SEMAPHORE semaphore)
{
  NUSE_Semaphore_Counter[semaphore] =
    NUSE_Semaphore_Initial_Value[semaphore];

  #if NUSE_BLOCKING_ENABLE

    NUSE_Semaphore_Blocking_Count[semaphore] = 0;

  #endif
}
```

The semaphore's counter (`NUSE_Semaphore_Counter[semaphore]`) is initialized to the user-specified value (`NUSE_Semaphore_Initial_Value[semaphore]`).

If task blocking is enabled, the semaphore's blocked task counter (`NUSE_Semaphore_Blocking_Count[semaphore]`) is set to zero.

Initializing event groups

Here is the complete code for `NUSE_Init_Event_Group()`:

```
void NUSE_Init_Event_Group(NUSE_EVENT_GROUP group)
{
  NUSE_Event_Group_Data[group] = 0;

  #if NUSE_BLOCKING_ENABLE

    NUSE_Event_Group_Blocking_Count[group] = 0;

  #endif
}
```

The event group's flags are cleared; that is, `NUSE_Event_Group_Data[group]` is set to zero.

If task blocking is enabled, the event group's blocked task counter (`NUSE_Event_Group_Blocking_Count[group]`) is set to zero.

Initializing timers

Here is the complete code for `NUSE_Init_Timer()`:

```
void NUSE_Init_Timer(NUSE_TIMER timer)
{
  NUSE_Timer_Status[timer] = FALSE;
```

```
    NUSE_Timer_Value[timer] = NUSE_Timer_Initial_Time[timer];
    NUSE_Timer_Expirations_Counter[timer] = 0;
}
```

The timer's status (`NUSE_Timer_Status[timer]`) is set to "unused"; that is, `FALSE`.

Its count-down value (`NUSE_Timer_Value[timer]`) is initialized to the user-specified value (`NUSE_Timer_Initial_Time[timer]`).

Its expiration counter (`NUSE_Timer_Expirations_Counter[timer]`) is set to zero.

Application code initialization

Once the Nucleus SE data structures have been initialized, there is the opportunity for code to be executed that performs initialization of the application prior to task execution. There are many possible uses for this capability:

Initialization of application data structures. Explicit assignments are easier to understand and debug than allowing automatic initialization of static variables.

Kernel object assignments. Given that all kernel objects are created statically at build time and are identified by index values, it may be useful to assign "ownership" or define the usage of these objects. This might be done using `#define` symbols, but, if there are multiple instances of tasks, the object indexes may be best assigned through global arrays (indexed by the task's ID).

Device initialization. This may be a good opportunity to set up any peripheral devices.

Obviously, many of these things could have been achieved before Nucleus SE initialization has been performed, but the advantage of locating the application initialization code here is that kernel services (API calls) may now be employed. For example, a queue or a mailbox may be preloaded with data to be processed when a task starts.

There is a limitation on which API calls are permitted: No action may be taken that would normally result in invocation of the scheduler—for example, task suspension/blocking. The global variable `NUSE_Task_State` has been set to `NUSE_STARTUP_CONTEXT` to reflect this restriction.

Starting the scheduler

Once initialization has been completed, it only remains to start the scheduler in order to commence execution of the application code—the tasks. The scheduler options and operation of the various types of scheduler were covered in detail in Chapter 5, *The Scheduler*, so only a brief summery is required here.

The key points in the sequence are:

Set the global variable `NUSE_Task_State` to `NUSE_TASK_CONTEXT`.

Select the index of the first task to be run. If support for initial task state is enabled, a search is performed for the first *Ready* task; otherwise, the value 0 is used.

The scheduler—`NUSE_Scheduler()`—is called.

Exactly what occurs in the last step depends on which scheduler type has been selected. For run to completion, the scheduling loop is entered and tasks are called in sequence. For other scheduler types, the context for the first task is loaded and control passed to the task.

18

Diagnostics and error checking

Error handling is unlikely to be a major feature of any operating system (OS) intended for embedded systems applications. This is an inevitable result of resource limitations—and all embedded systems have some kind of constraints. It is also logical, as only a limited number of embedded systems have the opportunity to behave like a desktop system—that is, offer the user an opportunity to decide what to do next in the event of some exceptional event.

In Nucleus SE there are broadly three types of error checking:

Facilities to "sanity check" the selected configuration—just to make sure that selected options are consistent,

Optionally included code to check runtime behavior, and

Specific application program interface (API) functions that facilitate the design of more robust code.

These will all be covered in this chapter along with some ideas about user-implemented diagnostics.

Configuration checks

Nucleus SE is designed to be very user-configurable so that it can be tailored to make the best use of available resources. This configurability is a challenge, as the number of options, and the interdependencies between them, is quite large. As has been described in many of the previous chapters, most user configuration of Nucleus SE is performed by setting #define constants in the file nuse_config.h.

In order to help identify configuration errors, a file—nuse_config_check.h—is included (i.e., by means of a #include into nuse_config.c), which performs a number of consistency checks on the #define symbols. Here is an extract from this file:

```
/*** Tasks and task control ***/

#if NUSE_TASK_NUMBER < 1 || NUSE_TASK_NUMBER > 16
  #error NUSE: invalid number of tasks - must be 1-16
#endif
```

Embedded RTOS Design. DOI: https://doi.org/10.1016/B978-0-12-822851-7.00018-7

```
#if NUSE_TASK_RELINQUISH && (NUSE_SCHEDULER_TYPE ==
NUSE_PRIORITY_SCHEDULER)
  #error NUSE: NUSE_Task_Relinquish() selected - not valid with
                          priority scheduler

#endif

#if NUSE_TASK_RESUME && !NUSE_SUSPEND_ENABLE
  #error NUSE: NUSE_Task_Resume() selected - task suspend not
                          enabled

#endif

#if NUSE_TASK_SUSPEND && !NUSE_SUSPEND_ENABLE
  #error NUSE: NUSE_Task_Suspend() selected - task suspend not
                          enabled

#endif

#if NUSE_INITIAL_TASK_STATE_SUPPORT && !NUSE_SUSPEND_ENABLE
  #error NUSE: Initial task state enabled - task suspend not
                          enabled
#endif


/*** Partition pools ***/

#if NUSE_PARTITION_POOL_NUMBER > 16
  #error NUSE: invalid number of partition pools - must be 0-16
#endif

#if NUSE_PARTITION_POOL_NUMBER == 0

  #if NUSE_PARTITION_ALLOCATE
    #error NUSE: NUSE_Partition_Allocate() enabled – no
                  partition pools configured
  #endif

  #if NUSE_PARTITION_DEALLOCATE
    #error NUSE: NUSE_Partition_Deallocate() enabled – no
                  partition pools configured
  #endif

  #if NUSE_PARTITION_POOL_INFORMATION
    #error NUSE: NUSE_Partition_Pool_Information() enabled –
                  no partition pools configured
  #endif

#endif
```

The checks performed include the following:

Verification that at least one, but no more than 16 tasks have been configured,

Confirmation that selected API functions are not inconsistent with the chosen scheduler or other options,

Verification that no more than 16 instances of other kernel objects has been specified,

Confirmation that API functions have not been selected for objects that are not instantiated at all,

Ensuring that API functions for signals and system time are not selected when these facilities have not been enabled, and

Verification of selected scheduler type and associated options.

In all cases, detection of an error results in the compilation of a #error statement. This normally results in the compilation being terminated with the specified message.

This file does not make it impossible to create an illogical configuration but renders it highly unlikely.

Application program interface parameter checking

Like Nucleus RTOS, Nucleus SE has the facility to optionally include code to verify API function call parameters at run time. Normally this would only be employed during initial debugging and testing as the memory and runtime overhead would be undesirable in production code.

Parameter checking is enabled by setting NUSE_API_PARAMETER_ CHECKING in nuse_config.h to TRUE. This enables the compilation of the required additional code. Here is an example of an API function's parameter checking:

```
STATUS NUSE_Mailbox_Send(NUSE_MAILBOX mailbox, ADDR *message,
                                                U8 suspend)
{
  STATUS return_value;

  #if NUSE_API_PARAMETER_CHECKING
    if (mailbox >= NUSE_MAILBOX_NUMBER)
    {
      return NUSE_INVALID_MAILBOX;
    }

    if (message == NULL)
    {
      return NUSE_INVALID_POINTER;
    }
```

```
#if NUSE_BLOCKING_ENABLE
  if ((suspend!=NUSE_NO_SUSPEND) &&
     (suspend!=NUSE_SUSPEND))
  {
    return NUSE_INVALID_SUSPEND;
  }
#else
  if (suspend!=NUSE_NO_SUSPEND)
  {
    return NUSE_INVALID_SUSPEND;
  }
#endif
#endif
```

This parameter checking may result in the return of an error code from the API function call. These are all negative values of the form NUSE_INVALID_xxx (e.g., NUSE_INVALID_POINTER)—a complete set of definitions is included in nuse_codes.h.

Extra application code (perhaps conditionally compiled) may be included to process these error values, but it would probably be better to use the data monitoring facilities of a modern embedded debugger to detect them.

Parameter checking introduces overheads in memory (the extra code) and run time performance, so its use is somewhat intrusive. As the full source code to Nucleus SE is available to the developer, checking and debugging can be done "by hand" on production code, if absolute precision is required.

Task stack checking

So long as the run-to-completion scheduler is not in use, Nucleus SE has a task stack checking facility available, which is similar to that provided in Nucleus RTOS, and provides an indication of remaining stack space. This API call—NUSE_Task_Check_Stack()—was described in detail in Chapter 6, *Tasks*. Some ideas on stack error checking are discussed in "User diagnostics" section later in this chapter.

Version information

Nucleus RTOS and Nucleus SE have an API function that simply returns version/release information about the kernel.

Nucleus RTOS API call

Service call prototype:

```
CHAR *NU_Release_Information(VOID);
```

Parameters:

None

Returns:

Pointer to NULL-terminated version string

Nucleus SE API call

This API call supports the key functionality of the Nucleus RTOS API.

Service call prototype:

```
char *NUSE_Release_Information(void);
```

Parameters:

None

Returns:

Pointer to NULL-terminated version string

Nucleus SE implementation of release information

The implementation of this API call is almost trivial. A pointer to the string constant `NUSE_Release_Info`, which is declared and initialized in `nuse_globals.c`, is returned.

This string takes the form Nucleus SE—`Xyymmdd`, where:

`x` is the release status: `A` = alpha, `B` = beta, and `R` = released;

`yy` is the year of the release;

`mm` is the month of release; and

`dd` is the day of release.

Compatibility with Nucleus RTOS

Nucleus RTOS includes an optional facility for maintaining a history log. The kernel records details of various system activities. API functions are provided to enable the application program to:

Enable/disable history saving,

Make a history entry, and

Retrieve a history entry.

This capability is not supported by Nucleus SE.

Nucleus RTOS also includes some error management macros that perform assertions and provide a means by which a user-defined fatal error function may be called. These are conditionally included in an OS build. Nucleus SE does not support this type of facility.

User diagnostics

So far in this chapter we have looked at the diagnostic and error checking facilities provided by Nucleus SE itself. This is now a good opportunity to consider how user-defined or application-oriented diagnostics may be implemented using the facilities provided by the kernel and/or applying our knowledge of its internal structure and implementation.

Application-specific diagnostics

Almost any application program could have additional code added to check its own integrity at run time. Using a multitasking kernel, having a specific task do this job is very convenient and straightforward. Obviously, diagnostics that are very specific to the application are not within the scope of this book, but we can consider some broad ideas.

Memory checks

The correct operation of memory is obviously critical to the integrity of any processor-based system. Equally obviously, a catastrophic failure would prevent any software, let alone diagnostics, from running at all. But there are situations when a degree of failure occurs and is a major concern but does not totally prevent code execution. Testing memory is quite a complex topic, which is way beyond the scope of this book, so I can only give some general ideas.

The two most common faults that can develop specifically in RAM are: "stuck bits"—where a bit has the value zero or one and cannot be changed; and "cross talk"—where adjacent bits interfere with one another. These can both be tested by writing and reading back appropriate test patterns to each RAM location in turn. Some testing can only really be performed on start-up, before even a stack is established; for example, a "moving ones" test, where each bit in memory is set to one and every other bit is checked to ensure that it is zero. Other byte-by-byte pattern testing can be performed on the fly, so long as you ensure that a context switch cannot occur, while the RAM location is corrupted. Using the Nucleus SE critical section delimiting macros `NUSE_CS_Enter()` and `NUSE_CS_Exit()`, which are defined in `nuse_types.h`, is straightforward and portable.

Various types of ROM are also subject to occasional failure, but there is limited checking that software can do. A checksum generated when code is built would be useful. This could be checked at start-up and perhaps also during runtime.

A failure in memory addressing logic can affect both ROM and RAM. A test specifically for this may be devised, but it is most likely to show up during the other tests described earlier.

Peripheral device checks

Outside of the CPU, peripheral circuitry may be subject to failure. This will vary greatly, of course, from one system to another, but it is very common for devices to have some means to verify their integrity with diagnostic software. For example, a communications line may have a loop-back mode, whereby any data written to it is immediately returned. If a device fails totally, it may stop responding to its address, so ensuring that an illegal-address trap handler is implemented makes sense.

Watchdog servicing

Embedded systems designers commonly include a "watchdog" circuit. This is a peripheral device that either interrupts the CPU and expects a prompt response, or (better) requires a periodic access initiated by the software. In either case, a common result of a watchdog "biting" is a system reset.

Using a watchdog effectively in a multitasking environment is challenging. Simply servicing it from one task only confirms that this particular task is functioning. One way around this is to implement a "supervisor task"—an example of this is described later in this chapter.

Stack overflow checking

Unless you select the run-to-completion scheduler, a Nucleus SE application will include a stack for each task. The integrity of these stacks is vital, but RAM is likely to be limited, so getting the size to be optimal is essential. It is possible, but very difficult, to statically predict the stack requirements for each task—it would need to have sufficient capacity for the demands of the most nested function call combined with the most stack-hungry interrupt service routine. A simpler approach is to perform exhaustive run-time testing.

There are broadly two approaches to stack verification. If a sophisticated embedded software debugger is in use, the stack boundaries may be monitored, and any transgressions detected. The location and size of Nucleus SE stacks are readily accessible via the global data structures in ROM: `NUSE_Task_Stack_Base[]` and `NUSE_Task_Stack_Size[]`.

The alternative is to perform run-time testing. The usual approach is to add "guard words" at the end of each stack—normally these would be the first location in each stack data area. These words are initialized to a recognizable nonzero value. Then, a diagnostic routine/task checks whether they have changed and takes appropriate action. The overwriting of a guard word does not mean the stack has actually overflowed but is an indication that it is about to do so; thus the software is likely to continue executing long enough to take corrective action or warn the user.

A supervisor task

Although Nucleus SE does not reserve any of the possible 16 tasks for its own use, the user may choose to dedicate one to diagnostics. This may be a low priority task, which just takes advantage of any "spare" CPU time; or it may be a high priority task that runs for a short time occasionally and thus ensures that diagnostics are always performed on a regular basis.

Here is how an example might function:

The supervisor task's signal flags are used to monitor the operation of six critical tasks in the system. These tasks each use a specific flag (bit 0 to bit 5) and are required to set it on a regular basis. The supervisor task clears all the flags and then suspends itself for a period of time. When it resumes, it expects all six tasks to have "checked in" by setting the appropriate flag; it seeks an exact match to the value `b00111111` (from `nuse_binary.h`). If this is satisfied, it clears the flags and suspends again. If not, it calls the critical error handling routine, which may perform system reset, for example.

An alternative implementation may have used an event flag group. This would make sense if signals were not used elsewhere in the application (as a RAM overhead would be incurred for all tasks) and even more so if event flags were employed for other purposes.

Tracing and profiling

Although many modern embedded software debuggers are quite customizable and could be made "RTOS aware," debugging a multithreaded application can still be challenging. A widely used approach is postexecution profiling, where the (RTOS) code is instrumented so that a detailed audit of its operations may be retrospectively analyzed. Typically, implementing such a facility involves two components:

First, additional code is added to the RTOS (i.e., it is "instrumented") to log activity. Typically, this will be enclosed in preprocessor directives to facilitate conditional compilation. This code stores a few bytes of information when a significant event occurs—like an API call or a context switch. This information would be things like:

The current address (PC),

Current task ID (index),

Index of any other object involved, and

A code indicating the operation that was performed.

Second, a task dedicated to emptying the buffer of profile information to external storage—typically a host computer.

The analysis of this captured data would also need some work, but this may be as simple as using an Excel spreadsheet.

19

Unimplemented facilities and compatibility

A key design requirement of Nucleus SE was a high level of compatibility with Mentor's flagship real-time operating system (RTOS) product, Nucleus RTOS. Nucleus SE has a subset of the functionality of Nucleus RTOS and I have highlighted this on all relevant occasions in this book, but, in this chapter, I will attempt to bring together all the key differences in one place. The intention is to give a quick reference for anyone planning migration between the two kernels or making a selection of a kernel for a specific project.

Apart from having limited or simplified functionality, as compared to Nucleus RTOS, Nucleus SE is also designed to be as memory-efficient as possible, with plenty of opportunities for tuning by the user. A key part of this strategy is scalable functionality. Many features of the kernel's functionality can be enabled or disabled, as required. Obviously, disabling functionality increases incompatibility with Nucleus RTOS for a given implementation.

In Nucleus RTOS, a system may be built with indefinite numbers of kernel objects—the only major limitation is the amount of available resources (i.e., memory). Nucleus SE has a limit of no more than 16 of each object type; a system may have 1–16 tasks and 0–16 of each other type of object (mailboxes, queues, etc.). Although this limit could be increased, it would take a considerable amount of work, as the ability to store an object index in a nibble (four bits) is widely exploited. Additionally, the priority scheduler is likely to become very inefficient if more than 16 tasks were permitted. An application that is seriously constrained by these limits is not really suited to Nucleus SE and Nucleus RTOS is likely to be a much better choice.

The scheduler

Like any modern real-time kernel, Nucleus RTOS has a very flexible scheduler, offering numerous priority levels (with an indefinite number of tasks on any given level); along with the

Embedded RTOS Design. DOI: https://doi.org/10.1016/B978-0-12-822851-7.00019-9

possibility for round robin and time slice scheduling. Nucleus SE is much simpler, offering four different schedulers, which must be selected at build time: run to completion, round robin, time slice, and priority. There is no option to combine the scheduling methods (i.e., there is no composite scheduler)—mixing priority with time slice, for example. Also, the priority scheduler only allows a single task at each priority level—there are only as many priority levels as there are tasks. A task's priority is fixed at build time, as is the time slice, if that option is used.

API calls

The application program interface (API) is the visible "face" of an operating system. It is unsurprising that it is here that the differences between Nucleus RTOS and Nucleus SE are most obvious.

Nucleus SE does not have the same API as Nucleus RTOS. However, its API is carefully designed to map very readily onto a subset of the Nucleus RTOS API. Licensees of Nucleus RTOS can use a "wrapper" (a header file containing `#define` macros) which render the mapping almost completely transparent.

Because the Nucleus SE API is a subset of Nucleus RTOS, there is the implication that some API calls are missing. This is true and an inevitable result of the design criteria of Nucleus SE. Some API calls would just be irrelevant, as they apply to nonexistent functionality; others are missing because of simplifications in the implementation of some kernel objects. This is all detailed in the following sections of this chapter.

Common application program interface functions

In Nucleus RTOS there are some API functions that are common across a number of different types of kernel object—or even across all of them. Some of these are also implemented in Nucleus SE— "reset" is a good example. Others are not applicable to the Nucleus SE implementation of kernel objects.

Create and delete

In Nucleus RTOS, all kernel objects are dynamic—they are created and deleted as required. Hence API calls are provided for this purpose. In Nucleus SE all objects are static—they are created at build time—so such API calls are not required.

Return of object pointers

The primary identifier (handle) used for kernel objects by Nucleus RTOS is a pointer to the object's control block, which is assigned when the object is created. There is, therefore, a set of API calls that return a list of pointers to objects of each type. Since Nucleus SE uses a simple index to identify a kernel object, such a call would be redundant. A program can interrogate the kernel to ascertain how many instances of a given object type are configured (using a call like `NUSE_Mailbox_Count()`); if this value is n, the indexes for the object type will range from zero to n-1.

Broadcast data

For a number of Nucleus RTOS kernel object types (notably mailbox, queue, and pipe), a "broadcast" API call is provided. This facilitates sending a data item to every task that is blocked on reading from the object. This capability was omitted from Nucleus SE for simplicity, as access to data in such objects is always gained in the context of the relevant task, which then frees the object; an additional flagging mechanism would be required in order to implement a broadcast.

Object-specific application program interface functions

Many kernel objects have API calls that are very specific to the particular object type and vary between Nucleus RTOS and Nucleus SE.

Tasks

Since the Nucleus RTOS scheduler is considerably more sophisticated than Nucleus SE's, a number of facilities provided by API functions are not required:

Change a task's preemption posture—not supported by Nucleus SE.

Change a task's priority—priorities are set at configuration time with Nucleus SE and cannot be changed.

Change a task's time slice—the time slice value is global for all tasks and fixed at configuration time in Nucleus SE.

Terminate a task—the *Terminated* task state is not supported by Nucleus SE.

Dynamic memory

As everything is created statically in Nucleus SE, dynamic memory is not supported (or required). Hence a number of specific API functions are not needed.

Signals

Nucleus RTOS supports signal handlers—routines that are run (rather like interrupt service routines) when a task's signals are modified. This capability was omitted from Nucleus SE and, hence, the API calls to control signals and register a signal handler are not required.

Interrupts

Nucleus SE takes a "hands off" attitude to interrupts, simply facilitating the option to perform some API calls from within an interrupt service routine. Hence, the set of Nucleus RTOS API calls that specify how the kernel processes interrupts are not required.

Diagnostics

Nucleus SE has very modest diagnostic facilities, in keeping with its "lean" design, being limited to (optional) parameter checking and the reporting of the product version code. Hence, Nucleus RTOS API calls associated with history logging and assertions are not implemented.

Drivers

Nucleus RTOS has a well-defined, formal driver structure, with a number of API functions associated with the management of drivers. Nucleus SE does not have such a structure, so the relevant API calls are not required.

API call functionality

Several of aspects of Nucleus SE's functionality, which are implemented in a simplified way, result in differences from Nucleus RTOS. A number of these impact the way API calls are used and the facilities available.

Timeouts

With Nucleus RTOS, there are many situations when an API call may optionally suspend a task pending the availability of a resource—the task is blocked. This suspension may be indefinite—that is, until the resource is available—or a timeout value may be specified. Nucleus SE offers blocking API calls as an option, but only indefinite suspension may be specified—that is, the call may only include **NUSE_SUSPEND** or **NUSE_NO_SUSPEND**, not a timeout value. This capability could be added to Nucleus SE in a reasonably straightforward manner.

Suspend order

When many types of object are created with Nucleus RTOS, the suspend order may be specified. This is the sequence in which a number of blocked tasks will be resumed as resources become available. Two options are available: first in first out, where tasks are resumed in the same order that they are blocked, or priority order, where the highest priority task is always resumed first. Nucleus SE does not offer this choice. Only priority order is implemented. Actually, the order is by task index, as this does not only apply to the priority scheduler, but also to round robin and time slice.

Object-specific functionality

In some cases, there is a change in functionality that is quite specific to a particular type of object.

Signal handlers

As mentioned earlier in this chapter, the implementation of signals in Nucleus SE does not support signal handling routines.

Application timer parameters

A timer has an initial duration and a restart duration and may optionally execute a user-specified function on expiration. This functionality is all supported in both Nucleus RTOS and Nucleus SE. However, Nucleus SE, unlike Nucleus RTOS, does not allow any of these parameters to be changed when a reset API call is made. Additionally, in Nucleus SE the entire support for expiration routines is optional.

Event flags

With Nucleus RTOS, there is an option to "consume" event flags. This means that flags that meet a task's criteria for a match are cleared. This functionality is not offered in Nucleus

SE as complexity is greatly increased by accommodating the possibility of multiple tasks' match criteria being met.

Data sizes

The two Nucleus SE design criteria of maintaining simplicity and minimizing memory usage result in a number of differences in the size of data items as compared with Nucleus RTOS. It should be noted that Nucleus RTOS commonly uses data of type **unsigned**, which is probably 32 bits; whereas Nucleus SE uses rationalized data types such as **U32**, **U16**, and **U8**.

Mailboxes

In Nucleus RTOS a mailbox carries a message consisting of four **unsigned** data items. In Nucleus SE a mailbox carries a single data item of type **ADDR**. My thinking is that a common use for mailboxes is the passing of an address (pointing to some data) between tasks.

Queues

In Nucleus RTOS a queue handles messages of one or more unsigned data elements; a queue may also be configured to handle messages of variable size. In Nucleus SE a queue handles messages that consist of a single data item of type **ADDR**. My thinking is that queues are used in a similar way to mailboxes. Also, in Nucleus RTOS the total size of the queue (i.e., total number of unsigned elements for which there is space) is specified as an **unsigned** value. In Nucleus SE this value is of type **U8**. A queue thus has smaller data capacity.

Pipes

In Nucleus RTOS a pipe handles messages of one or more bytes; a pipe may also be configured to handle messages of variable size. In Nucleus SE a pipe handles messages that consist of one or more data items of type **U8**. The message size is set at configuration time for each pipe. Also, in Nucleus RTOS the total size of the pipe (i.e., total number of bytes for which there is space) is specified as an **unsigned** value. In Nucleus SE this value is of type **U8** and represents the number of messages (in the **NUSE_Pipe_Information()** API call). A pipe thus has smaller data capacity.

Event flag groups

In Nucleus RTOS an event flag group contains 32 flags; in Nucleus SE this is reduced to eight. This size was chosen as likely target processors for Nucleus SE efficiently handle 8-bit data. It would not be difficult to change Nucleus SE to handle event flag groups of a different size.

Signals

In Nucleus RTOS every task has a set of 32 signal flags. In Nucleus SE signals are optional and each task has a set of just eight flags. This size was chosen as likely target processors for Nucleus SE efficiently handle 8-bit data. It would not be difficult to change Nucleus SE to handle signal flag sets of a different size.

Memory partitions

In Nucleus RTOS the specification of the number and size of partitions are both **unsigned** parameters. In Nucleus SE the number of partitions is a parameter of type **U8** and the partition size is **U16**. This implies some limitations on the partition and pool size.

Timers

In Nucleus RTOS timers (both application timers and task sleep) handle values of type **unsigned**. In Nucleus SE they are of type **U16**. This type was chosen as likely target processors for Nucleus SE efficiently handle 16-bit data (and 8 bits would not be enough to be useful). It would not be difficult to change Nucleus SE to handle timers of a different size.

20

Using Nucleus SE

So far in this book we have looked in detail at all the facilities that Nucleus SE has to offer. Now it is time to see how to use it for a real embedded software application.

What is Nucleus SE?

We know that Nucleus SE is a real-time kernel, but it is important to understand how that fits in with the rest of an application. "Fitting in" is exactly what it does, because, unlike a desktop operating system (OS), such as Windows, you do not really run an application on Nucleus SE; the kernel is simply part of the application software that runs on the embedded device. This is very commonly the case with real-time operating systems (RTOSs).

At the highest level, a conventional embedded application is some code, that is run when the CPU is reset. This initializes the hardware and software environment and then calls the `main()` function, which is the start of the application code. What is different, when using Nucleus SE (or many other similar kernels) is that the `main()` function is supplied as part of the kernel code. This function simply initializes all the kernel data structures, and then it calls the scheduler, which results in the application code (tasks) being run. The user may wish to add any application-specific initialization to the `main()` function.

Nucleus SE also includes a selection of functions—the application program interface (API)—that provide a range of facilities like inter-task communication and synchronization, timing, memory allocation, etc. All the API functions have been described earlier in this book.

All the Nucleus SE software is provided as (mostly C language) source code. Conditional compilation is used to configure the code to the requirements of a specific application. This is described in detail in the *Configuring a Nucleus SE Application* section later in this chapter.

When the code has been compiled, the resulting Nucleus SE object modules are linked with those of the application code to

Embedded RTOS Design. DOI: https://doi.org/10.1016/B978-0-12-822851-7.00020-5

result in a single binary image, which would normally be placed in flash memory in the embedded device. The result of such a static link is that all symbolic information may remain available—both from the application code and the kernel. This is a useful aid to debugging, but care is needed to avoid misuse of Nucleus SE data.

CPU and tool support

Since Nucleus SE is supplied in source code, the intention is for it to be portable. However, it is impossible for code that works at such a low level (i.e., with a scheduler other than run to completion, where context switches are required) to be totally free of assembly language. But I have minimized this, and very little low-level coding should be needed to port to a new CPU. Using a new development toolkit (compiler, assembler, linker, etc.) may also raise portability issues.

Configuring a Nucleus SE application

The key to using Nucleus SE efficiently is getting the configuration right. This may appear complex but is actually quite logical and just needs to be approached systematically. Almost all the configuration is performed by editing two files: `nuse_config.h` and `nuse_config.c`.

Setting up nuse_config.h

This file is simply a list of `#define` symbols, which are set to appropriate values to specify the desired kernel configuration. In the default `nuse_config.h` file, all the symbols are present, but set to a minimal configuration.

Object counts

The number of each kernel object type is set by assigning values to symbols of the form `NUSE_SEMAPHORE_NUMBER`. For most objects, the value can be 0—15. Tasks are the exception, as there must be at least one. Signals are not really objects in their own right, as they are associated with tasks and enabled by setting `NUSE_SIGNAL_SUPPORT` to `TRUE`.

API function enables

Each Nucleus SE API function may be individually enabled by setting a symbol with the same name as the function—for example, NUSE_PIPE_JAM—to TRUE. This results in the code for the function being included in the application.

Scheduler selection and settings

Nucleus SE supports four types of scheduler, as described in detail in Chapter 5, *The Scheduler*. The type of scheduler to be used for an application may be specified by setting NUSE_SCHEDULER_TYPE to one of these values: NUSE_RUN_TO_COMPLETION_SCHEDULER, NUSE_TIME_SLICE_SCHEDULER, NUSE_ROUND_ROBIN_SCHEDULER, or NUSE_PRIORITY_SCHEDULER.

Other aspects of the scheduler may also be set up:

NUSE_TIME_SLICE_TICKS specifies the number of ticks per slot for the time slice scheduler. If another scheduler is in use, this must be set to 0.

NUSE_SCHEDULE_COUNT_SUPPORT can be set to TRUE or FALSE to enable/disable the task scheduler counting mechanism.

NUSE_SUSPEND_ENABLE facilitates support for task suspend. If this is set to FALSE, tasks can never be suspended and are always ready to be run. If the priority scheduler is used, this option must be set to TRUE.

NUSE_BLOCKING_ENABLE enables task blocking (suspend) on many API functions. This means that a call to such a function may result in suspension of the calling task, pending the availability of a resource. Selecting this option requires that NUSE_SUSPEND_ENABLE is also set to TRUE.

Other options

A few other options may be set to TRUE or FALSE to enable/disable other kernel functionality:

NUSE_API_PARAMETER_CHECKING enables inclusion of code to verify API function call parameters and would normally be set during debugging.

NUSE_INITIAL_TASK_STATE_SUPPORT enables the initial state of all tasks to be specified as NUSE_READY or NUSE_PURE_SUSPEND. If this facility is disabled, all tasks start with the status NUSE_READY.

NUSE_TIMER_EXPIRATION_ROUTINE_SUPPORT enables support for a call to a function to be made when an application timer expires. If it is disabled, no action is taken on timer expiry.

NUSE_SYSTEM_TIME_SUPPORT enables the system tick clock.

NUSE_INCLUDE_EVERYTHING is an option to include as much as possible into the Nucleus SE configuration. It results in the activation of all optional functionality and every API function for objects that have been configured. It is used as shorthand to create a Nucleus SE configuration to exercise a new port of the kernel code.

Setting up nuse_config.c

After specifying the kernel configuration in nuse_config.h, various ROM-based data structures need to be initialized. This is done in nuse_config.c. Definition of the data structures is controlled by conditional compilation, so they are all present in the default copy of nuse_config.c.

Task data

The array NUSE_Task_Start_Address[] should be initialized with the start addresses for each task—this is normally just a list of function names (without the parentheses). Prototypes for the task entry functions must also be visible. In the default file, a single task is configured with the name NUSE_Idle_Task()—this may be replaced with an application task.

Unless the run-to-completion scheduler is in use, each task requires its own stack. An array in RAM much be created for each task's stack. The arrays should be of type ADDR and the address of each one placed in NUSE_Task_Stack_Base[]. Estimating the size of arrays is difficult and best done by measurement—see the *Debugging a Nucleus SE Application* section later in this chapter. The size of each array (i.e., stack size in words) should be placed in NUSE_Task_Stack_Size[].

If the facility to specify initial task status has been enabled (using NUSE_INITIAL_TASK_STATE_SUPPORT), the array NUSE_Task_Initial_State[] should be initialized to NUSE_READY or NUSE_PURE_SUSPEND.

Partition pool data

If any partition pools are configured, an array (of type U8) needs to be defined in RAM for each one. The size of these arrays is computed thus: [number of partitions × (partition size + 1)]. The addresses of these arrays (i.e., just their names) should be assigned to the appropriate elements of `NUSE_Partition_Pool_Data_Address[]`. For each pool, the number of partitions and the size of those partitions should be assigned into `NUSE_Partition_Pool_Partition_Number[]` and `NUSE_Partition_Pool_Partition_Size[]`, respectively.

Queue data

If any queues are configured, an array (of type `ADDR`) needs to be defined in RAM for each one. The size of these arrays is simply the number of elements required in each queue. The addresses of these arrays (i.e., just their names) should be assigned to the appropriate elements of `NUSE_Queue_Data[]`. For each queue, its size should be assigned into the appropriate element of `NUSE_Queue_Size[]`.

Pipe data

If any pipes are configured, an array (of type `U8`) needs to be defined in RAM for each one. The size of these arrays is computed thus: (pipe size × pipe message size). The addresses of these arrays (i.e., just their names) should be assigned to the appropriate elements of `NUSE_Pipe_Data[]`. For each pipe, its size and message size should be assigned into the appropriate elements of `NUSE_Pipe_Size[]` and `NUSE_Pipe_Message_Size[]`, respectively.

Semaphore data

If any semaphores are configured, the array `NUSE_Semaphore_Initial_Value[]` needs to be initialized to the downcounter start values.

Application timer data

If any timers are configured, the array `NUSE_Timer_Initial_Time[]` needs to be initialized to the start values for the counters. Also, `NUSE_Timer_Reschedule_Time[]` needs to be set to the restart values. These are counter values to be used after the initial sequence expires. If the restart values are set to 0, the counter stops after one cycle.

If support for timer expiration routines is configured (by setting `NUSE_TIMER_EXPIRATION_ROUTINE_SUPPORT` to `TRUE`), two more arrays need to be initialized. The addresses of the expiration routines (just a list of function names without the parentheses) should be assigned to `NUSE_Timer_Expiration_Routine_Address[]`.

The array `NUSE_Timer_Expiration_Routine_Parameter[]` should be initialized to the expiration routine parameter values.

Which application program interface?

All OSes have an API of some kind. Nucleus SE is no exception and the function calls that make up its API have been covered extensively in this book.

So, it would seem obvious that, when writing an application that incorporates Nucleus SE you would use its API as described. This may not always be the case.

For many users, the Nucleus SE API will be new and may be their first experience of using an OS's API and, since it is fairly simple and straightforward, it may make a very good introduction to the topic. In this case, the way forward is clear.

For some other users, an alternative API might be attractive. There are three obvious situations when this might be the case:

1. The Nucleus SE application is just part of a system where other OS(es) is/are in use for other components. Having portability of code and, more importantly, expertise between the OSes in use is very attractive.
2. The user has extensive experience of another OS's API. Reusing that expertise is very desirable.
3. The user wishes the reuse code, which was written for the API of some other OS. Recoding to change the API calls is possible, but time-consuming.

Since the complete source code to Nucleus SE is provided, it would be entirely possible to edit each API function to appear to be the same as its equivalent in another OS. However, this would be time-consuming and ultimately unproductive process. A better approach is to write a "wrapper." This can be done in various ways, but the simplest is simply a header (`#include`) file which contains a series of `#define` macros that map from a "foreign" API to Nucleus SE's API.

A wrapper that maps (a subset of) the Nucleus RTOS API onto Nucleus SE is included in the distribution. This can be used by developers with experience of Nucleus RTOS or where migration to this RTOS is a future possibility. This wrapper also serves as an example to aid the development of a wrapper for another RTOS.

Debugging a Nucleus SE application

Writing an embedded application using a multitasking kernel is a challenge. Verifying that this code works, and the identification of bugs can be very problematic. Although it is just code executing on a processor, the apparent concurrent execution of multiple tasks means that focusing on a particular thread of execution is not easy. This is compounded when code is shared between tasks; worst of all when two tasks use entirely the same code (but operate on different data, obviously). An additional issue is the untangling of data structures, which are used to implement kernel objects, in order to see the information in a meaningful fashion.

To debug an application built with Nucleus SE does not require any special libraries or other facilities. All the kernel source code is there and may be "visible" to a debugger. Hence all the symbolic information is available for use and interrogation. Any modern debugging tool may be employed to work on a Nucleus SE-based application.

Using a debugger

Debugging tools that are designed specifically for embedded applications have been with us for more than 30 years now and have, hence, become very sophisticated. The key characteristic of an embedded application, as compared with a desktop program, is that every embedded system is different (but one PC looks very much like every other). The trick with a good embedded debugger is for it to be flexible and customizable enough to accommodate such variability in requirements from one user to another. The customizability of a debugger is manifest in various forms, but there is generally some scripting capability. It is this facility in particular which may be exploited to make a debugger perform well with a kernel-based application. I will review some of the possibilities here.

It is important to note that a debugger is typically a family of tools, not just a single program. A debugger may have different modes of operation whereby it can assist with development of code on a simulated target or with real target hardware.

Task aware breakpoints

If code is shared by multiple tasks in an application, it makes conventional debugging using breakpoints rather confusing. It is likely that you will only want the code to stop when a breakpoint is reached in the context of the particular task you are endeavoring to debug. What you need is a task aware breakpoint.

Fortunately, the scripting facilities of a modern debugger and the visibility of Nucleus SE symbols make implementing task aware breakpoints quite straightforward. All that is needed is a simple script that is attached to a breakpoint that you wish to make "task aware." This script would take a parameter, which is the index (ID) of the task in which you are interested. The script would simply compare this value with the index of the currently running task (in `NUSE_Task_Active`). If the values match, execution is halted; if they are different, execution is allowed to continue. It is fair to point out that the execution of this script will have some effect on the real-time profile of the application, but, unless it is in a loop where the script is likely to be executed very frequently, that effect will be minimal.

Kernel object information

An obvious need, while debugging a Nucleus SE-based application, is the ability to find out about kernel objects—what are their characteristics and current status. This is a matter of answering questions like: "How big is a queue and how many messages are in it?"

One way to facilitate this is to add some additional debug code to your application, which can make use of the "information" API calls—such as `NUSE_Queue_Information()`. This, of course, means that your application contains extra code, which will not be needed after deployment. Using a `#define` symbol to switch this code in and out, using conditional compilation, would be a sensible solution.

Some debuggers can perform a target function call—that is, directly call the information API function. This gets around the need to add extra code, except that the API function must have been configured for the debugger to make use of it.

An alternative approach, that is more flexible, but less "future proof" is to directly access the kernel object's data structures. This is probably best done using the debugger's scripting capabilities. In our example, the size of the queue may be obtained from `NUSE_Queue_Size[]` and their current usage from `NUSE_Queue_Items[]`. Furthermore, using the address of the queue's data area (from `NUSE_Queue_Data[]`) and the head/tail pointers (`NUSE_Queue_Head[]` and `NUSE_Queue_Tail[]`), the queued messages may be displayed.

API call return values

Many API functions return a status value, which indicates whether the call was successful. It would be useful to monitor these values and flag instances where they are not `NUSE_SUCCESS` (which has the value zero). Since this monitoring is for debug purposes only, conditional compilation is in order. The definition of a global RAM variable (`NUSE_API_Call_Status`, say) can be conditionally compiled (under the control of a `#define` symbol). Then the assignment part of the API calls (i.e., the `NUSE_API_Call_Status =`) can be similarly conditionally compiled. For example, for debugging purposes, a call that would normally look like this:

```
NUSE_Mailbox_Send(mbox, msg, NUSE_SUSPEND);
```

becomes

```
NUSE_API_Call_Status = NUSE_Mailbox_Send(mbox, msg, NUSE_SUSPEND);
```

If task blocking is enabled, many API function calls can only return success or an indication that the object has been reset. However, if API parameter checking is enabled, a variety of other return values are possible.

Task stack sizing and overflow

The topic of stack overflow protection was discussed in Chapter 18, Diagnostics and Error Checking. During debugging, there are a couple of other possibilities:

A stack memory area could be filled with a characteristic value—something other than all ones or all zeroes. The debugger can then be used to watch the memory locations and the extent to which the values get changed indicates the extent of stack usage. If all the memory locations have been changed, it does not necessarily mean that the stack has overflowed but may mean that the stack is only just large enough, which is fragile. It should be enlarged and be subject to further testing.

As discussed, when implementing diagnostics in Chapter 18, *Diagnostics and Error Checking*, addition locations—"guard words"—may be located at either end of the stack memory area. The debugger can be used to monitor access to these words, as any write would indicate stack underflow or overflow.

Nucleus SE configuration checklist

Because Nucleus SE is designed to be very flexible and customizable to accommodate the precise needs of the application, a significant amount of configuration is required. This is why this whole chapter is essentially dedicated to the topic. To make it easy to ensure that everything is covered, here is a checklist of all the key steps involved in build a Nucleus SE-based embedded application:

1. *Obtain Nucleus SE*—Although almost all the code of Nucleus SE has been published in this book, the next chapter will tell you how to obtain it in a more immediately usable form.
2. *Consider CPU/tool support*—There may be a need to rewrite the assembly language parts and redraft build scripts.
3. *Build the simple demo*—This verifies that you have all the components and the tools are compatible.
4. *Plan your task structure*—How many tasks and what do they do. Set up their start addresses and stack sizes. You can adjust this later, of course. You may have up to 16 tasks in an application.
5. *Application initialization*—Do you need to add any code to `main()`?
6. *Select scheduler*—You have four to choose from and this might be changed later.
7. *Verify timer interrupt vector*—If you have a timer.
8. *Verify the context switch trap vector*—If you are not using the run-to-completion scheduler.
9. *Signals*—Enable support for signals if you are going to use them.
10. *System clock*—Enable the clock if you need it.
11. *Kernel object counts*—Determine how many of each kind of object you need. You can change this later. Maximum of 16 objects of each type.
12. *Kernel object ROM*—Initialize the ROM data for each type of object that you are using.
13. *RAM data*—Set up RAM data space for objects that need it (queues, pipes, and partition pools).
14. *API enables*—Enable all the API calls that you need.

21

Nucleus SE reference information

In this, the final chapter, I am summarizing and bringing together some reference material that has appeared in previous chapters. First is a listing of all the Nucleus SE API calls and then some details of all the symbols and data structures used in Nucleus SE.

Alphabetic API function list

Nucleus SE has a total of 49 application program interface (API) functions. These are all described in the relevant chapters and a summary is included here, in alphabetical order, for reference.

The naming convention of the functions, with a single exception, is to start with the prefix `NUSE_`, then have a category (typically, a kernel object type, like semaphore, for example), an underscore, then a description of the function's purpose. This convention results in an alphabetic listing, which also groups the functions in a logical way. The single exception to the naming scheme is `NUSE_Release_Information()`, which uniquely does not relate to a type of kernel object.

NUSE_Clock_Retrieve()

Returns the current system time value (see Chapter 14: *System Time*).

Service call prototype:

```
U32 NUSE_Clock_Retrieve(void);
```

Parameters:

None

Returns:

The current system time value

Embedded RTOS Design. DOI: https://doi.org/10.1016/B978-0-12-822851-7.00021-7

NUSE_Clock_Set()

Sets a new system time value.

Service call prototype:

```
void NUSE_Clock_Set(U32 new_value);
```

Parameters:

new_value—the value to which the system time is to be set

Returns:

None

NUSE_Event_Group_Count()

Obtains the number of event flag groups configured in the application (see Chapter 9: *Event Flag Groups*).

Service call prototype:

```
U8 NUSE_Event_Group_Count(void);
```

Parameters:

None

Returns:

The number of configured event flag groups in the application

NUSE_Event_Group_Information()

Obtains information about a specific event flag group.

Service call prototype:

```
STATUS NUSE_Event_Group_Information(
NUSE_EVENT_GROUP group,
U8 *event_flags,
U8 *tasks_waiting,
NUSE_TASK *first_task);
```

Parameters:

group—the index of the event flag group about which information is being requested;

event_flags—a pointer to a variable, which will receive the current value of the specified event flag group;

tasks_waiting—a pointer to a variable which will receive the number of tasks suspended on this event flag group (nothing returned if task suspend is disabled);

first_task—a pointer to a variable of type NUSE_TASK which will receive the index of the first suspended task (nothing returned if task suspend is disabled).

Returns:

NUSE_SUCCESS—The call was completed successfully.

NUSE_INVALID_GROUP—The event flag group index is not valid.

NUSE_INVALID_POINTER—One or more of the pointer parameters is invalid.

NUSE_Event_Group_Retrieve()

Obtains a specific pattern of event flags from an event flag group.

Service call prototype:

```
STATUS NUSE_Event_Group_Retrieve(
NUSE_EVENT_GROUP group,
U8 requested_events,
OPTION operation,
U8 *retrieved_events,
U8 suspend);
```

Parameters:

group—the index (ID) of the event flag group to be utilized;

requested_events—a bit pattern representing the flags to be retrieved;

operation—specification for whether all or just some of the requested events need to be present; may be NUSE_OR (some flags) or NUSE_AND (all flags);

retrieved_events—a pointer to storage for the actual value of the retrieved event flags (same value as requested_events if operation was NUSE_AND);

suspend—specification for task suspend; may be NUSE_NO_SUSPEND or NUSE_SUSPEND.

Returns:

NUSE_SUCCESS—The call was completed successfully.

NUSE_NOT_PRESENT—Specified operation did not retrieve events (none present for NUSE_OR; not all present for NUSE_AND).

`NUSE_INVALID_GROUP`—The event flag group index is invalid.

`NUSE_INVALID_OPERATION`—The specified operation was not `NUSE_OR` or `NUSE_AND`.

`NUSE_INVALID_POINTER`—The pointer to retrieved event flag storage is `NULL`.

`NUSE_INVALID_SUSPEND`—Suspend was attempted from a nontask thread or when blocking API calls were not enabled.

NUSE_Event_Group_Set()

Sets or clears a specific pattern of event flags from an event flag group.

Service call prototype:

```
STATUS NUSE_Event_Group_Set(
NUSE_EVENT_GROUP group,
U8 event_flags,
OPTION operation);
```

Parameters:

`group`—the index (ID) of the event group in which flags are to be set/cleared;

`event_flags`—the bit value of the pattern of flags to be operated on;

`operation`—the operation to be performed; may be `NUSE_OR` (to set flags) or `NUSE_AND` (to clear flags).

Returns:

`NUSE_SUCCESS`—The call was completed successfully.

`NUSE_INVALID_GROUP`—The event flag group index is invalid.

`NUSE_INVALID_OPERATION`—The specified operation was not `NUSE_OR` or `NUSE_AND`.

NUSE_Mailbox_Count()

Obtains the number of mailboxes configured in the application (see Chapter 11: *Mailboxes*).

Service call prototype

```
U8 NUSE_Mailbox_Count(void);
```

Parameters:

> None

Returns:

> The number of configured mailboxes in the application

NUSE_Mailbox_Information()

Obtains information about a specific mailbox.

Service call prototype:

```
STATUS NUSE_Mailbox_Information(
NUSE_MAILBOX mailbox,
U8 *message_present,
U8 *tasks_waiting,
NUSE_TASK *first_task);
```

Parameters:

> `mailbox`—the index of the mailbox about which information is being requested;
>
> `message_present`—a pointer to a variable, which will receive a value of TRUE or FALSE depending on whether the mailbox is full or not;
>
> `tasks_waiting`—a pointer to a variable, which will receive the number of tasks suspended on this mailbox (nothing returned if task suspend is disabled);
>
> `first_task`—a pointer to a variable of type NUSE_TASK, which will receive the index of the first suspended task (nothing returned if task suspend is disabled).

Returns:

> `NUSE_SUCCESS`—The call was completed successfully.
>
> `NUSE_INVALID_MAILBOX`—The mailbox index is not valid.
>
> `NUSE_INVALID_POINTER`—One or more of the pointer parameters is invalid.

NUSE_Mailbox_Receive()

Reads data from a specified mailbox.

Service call prototype:

```
STATUS NUSE_Mailbox_Receive(
NUSE_MAILBOX mailbox,
```

```
ADDR *message,
U8 suspend);
```

Parameters:

> `mailbox`—the index (ID) of the mailbox to be utilized;
>
> `message`—a pointer to storage for the message to be received, which is a single variable of type `ADDR`;
>
> `suspend`—specification for task suspend; may be `NUSE_NO_SUSPEND` or `NUSE_SUSPEND`.

Returns:

> `NUSE_SUCCESS`—The call was completed successfully.
>
> `NUSE_INVALID_MAILBOX`—The mailbox index is invalid.
>
> `NUSE_INVALID_POINTER`—The message pointer is `NULL`.
>
> `NUSE_INVALID_SUSPEND`—Suspend was attempted from a nontask thread or when blocking API calls were not enabled.
>
> `NUSE_MAILBOX_EMPTY`—The mailbox is empty and suspend was not specified.
>
> `NUSE_MAILBOX_WAS_RESET`—The mailbox was reset, while the task was suspended.

NUSE_Mailbox_Reset()

Restores a mailbox to its initial, unused state.

Service call prototype:

```
STATUS NUSE_Mailbox_Reset(NUSE_MAILBOX mailbox);
```

Parameters:

> `mailbox`—the index (ID) of the mailbox to be reset.

Returns:

> `NUSE_SUCCESS`—The call was completed successfully.
>
> `NUSE_INVALID_MAILBOX`—The mailbox index is not valid.

NUSE_Mailbox_Send()

Writes data to a specified mailbox.

Service call prototype:

```
STATUS NUSE_Mailbox_Send(
```

```
NUSE_MAILBOX mailbox,
ADDR *message,
U8 suspend);
```

Parameters:

> `mailbox`—the index (ID) of the mailbox to be utilized;
>
> `message`—a pointer to the message to be sent, which is a single variable of type `ADDR`;
>
> `suspend`—specification for task suspend; may be `NUSE_NO_SUSPEND` or `NUSE_SUSPEND`.

Returns:

> `NUSE_SUCCESS`—The call was completed successfully.
>
> `NUSE_INVALID_MAILBOX`—The mailbox index is invalid.
>
> `NUSE_INVALID_POINTER`—The message pointer is `NULL`.
>
> `NUSE_INVALID_SUSPEND`—Suspend was attempted from a nontask thread or when blocking API calls were not enabled.
>
> `NUSE_MAILBOX_FULL`—The mailbox is full and suspend was not specified.
>
> `NUSE_MAILBOX_WAS_RESET`—The mailbox was reset, while the task was suspended.

NUSE_Partition_Allocate()

Allocates a memory partition in a specified pool (see Chapter 7: *Partition Memory*).

Service call prototype:

```
STATUS NUSE_Partition_Allocate(
NUSE_PARTITION_POOL pool,
ADDR *return_pointer,
U8 suspend);
```

Parameters:

> `pool`—the index (ID) of the partition pool to be utilized;
>
> `return_pointer`—a pointer to a variable of type `ADDR`, which will receive the address of the allocated partition;
>
> `suspend`—specification for task suspend; may be `NUSE_NO_SUSPEND` or `NUSE_SUSPEND`.

Returns:

>NUSE_SUCCESS—The call was completed successfully.

>NUSE_NO_PARTITION—No partitions are available.

>NUSE_INVALID_POOL—The partition pool index is invalid.

>NUSE_INVALID_POINTER—The data return pointer is NULL.

>NUSE_INVALID_SUSPEND—Suspend was attempted from a nontask thread or when blocking API calls were not enabled.

NUSE_Partition_Deallocate()

Deallocates a memory partition at a specified address.

Service call prototype:

>STATUS NUSE_Partition_Deallocate(ADDR partition);

Parameters:

>partition—pointer to the data area (as returned by NUSE_Partition_Allocate()) of the partition to be deallocated.

Returns:

>NUSE_SUCCESS—The call was completed successfully.

>NUSE_INVALID_POINTER—The partition pointer is NULL or does not appear to point to a valid, in-use partition.

NUSE_Partition_Pool_Count()

Obtains the number of partition pools configured in the application.

Service call prototype:

>U8 NUSE_Partition_Pool_Count(void);

Parameters:

>None

Returns:

>The number of configured partition pools in the application

NUSE_Partition_Pool_Information()

Obtains information about a specific partition pool.

Service call prototype

```
STATUS NUSE_Partition_Pool_Information(
NUSE_PARTITION_POOL pool,
ADDR *start_address,
U32 *pool_size,
U16 *partition_size,
U8 *available,
U8 *allocated,
U8 *tasks_waiting,
NUSE_TASK *first_task);
```

Parameters:

pool—the index of the partition pool about which information is being requested;

start_address—a pointer to a variable, which will receive a pointer to the start of the partition pool data area;

pool_size—a pointer to a variable, which will receive the size of the partition pool (in bytes);

partition_size—a pointer to a variable, which will receive the size of partitions in this pool;

available—a pointer to a variable, which will receive the number of partitions currently available in this pool;

allocated—a pointer to a variable, which will receive the number of partitions currently in use in this pool;

tasks_waiting—a pointer to a variable, which will receive the number of tasks suspended on this partition pool (nothing returned if task suspend is disabled);

first_task—a pointer to a variable of type NUSE_TASK, which will receive the index of the first suspended task (nothing returned if task suspend is disabled).

Returns:

NUSE_SUCCESS—The call was completed successfully

NUSE_INVALID_POOL—The partition pool index is not valid.

NUSE_INVALID_POINTER—One or more of the pointer parameters is invalid.

NUSE_Pipe_Count()

Obtains the number of pipes configured in the application (see Chapter 13: *Pipes*).

Service call prototype:

```
U8 NUSE_Pipe_Count(void);
```

Parameters:

None

Returns:

The number of configured pipes in the application

NUSE_Pipe_Information()

Obtains information about a specific pipe.

Service call prototype:

```
STATUS NUSE_Pipe_Information(NUSE_PIPE pipe,
ADDR *start_address,
U8 *pipe_size,
U8 *available,
U8 *messages,
U8 *message_size,
U8 *tasks_waiting,
NUSE_TASK *first_task);
```

Parameters:

`pipe`—the index of the pipe about which information is being requested;

`start_address`—a pointer to a variable of type `ADDR`, which will receive the address of the start of the pipe's data area;

`pipe_size`—a pointer to a variable of type `U8`, which will receive the total number of messages for which the pipe has capacity;

`available`—a pointer to a variable of type `U8`, which will receive the number of messages for which the pipe has currently remaining capacity;

`messages`—a pointer to a variable of type `U8`, which will receive the number of messages currently in the pipe;

`message_size`—a pointer to a variable of type `U8`, which will receive the size of messages handled by this pipe;

tasks_waiting—a pointer to a variable of type u8, which will receive the number of tasks suspended on this pipe (nothing returned if task suspend is disabled);

first_task—a pointer to a variable of type NUSE_TASK, which will receive the index of the first suspended task (nothing returned if task suspend is disabled).

Returns:

NUSE_SUCCESS—The call was completed successfully.

NUSE_INVALID_PIPE—The pipe index is not valid.

NUSE_INVALID_POINTER—One or more of the pointer parameters is invalid.

NUSE_Pipe_Jam()

Writes data to the front of a specified pipe.

Service call prototype:

```
STATUS NUSE_Pipe_Jam(
NUSE_PIPE pipe,
U8 *message,
U8 suspend);
```

Parameters:

pipe—the index (ID) of the pipe to be utilized;

message—a pointer to the message to be sent, which is a sequence of bytes as long as the configured message size of the pipe;

suspend—specification for task suspend; may be NUSE_NO_SUSPEND or NUSE_SUSPEND.

Returns:

NUSE_SUCCESS—The call was completed successfully.

NUSE_INVALID_PIPE—The pipe index is invalid.

NUSE_INVALID_POINTER—The message pointer is NULL.

NUSE_INVALID_SUSPEND—Suspend was attempted from a non-task thread or when blocking API calls were not enabled.

NUSE_PIPE_FULL—The pipe is full and suspend was not specified.

NUSE_PIPE_WAS_RESET—The pipe was reset while the task was suspended.

NUSE_Pipe_Receive()

Reads data from a specified pipe.

Service call prototype:

```
STATUS NUSE_Pipe_Receive(
NUSE_PIPE pipe,
U8 *message,
U8 suspend);
```

Parameters:

pipe—the index (ID) of the pipe to be utilized;

message—a pointer to storage for the message to be received, which is a sequence of bytes as long as the configured message size of the pipe;

suspend—specification for task suspend; may be NUSE_NO_SUSPEND or NUSE_SUSPEND.

Returns:

NUSE_SUCCESS—The call was completed successfully.

NUSE_INVALID_PIPE—The pipe index is invalid.

NUSE_INVALID_POINTER—The message pointer is NULL.

NUSE_INVALID_SUSPEND—Suspend was attempted from a non-task thread or when blocking API calls were not enabled.

NUSE_PIPE_EMPTY—The pipe is empty and suspend was not specified.

NUSE_PIPE_WAS_RESET—The pipe was reset, while the task was suspended.

NUSE_Pipe_Reset()

Restores a pipe to its initial, unused state.

Service call prototype:

```
STATUS NUSE_Pipe_Reset(NUSE_PIPE pipe);
```

Parameters:

pipe—the index (ID) of the pipe to be reset.

Returns:

NUSE_SUCCESS—The call was completed successfully.

NUSE_INVALID_PIPE—The pipe index is not valid.

NUSE_Pipe_Send()

Writes data to the back of a specified pipe.

Service call prototype:

```
STATUS NUSE_Pipe_Send(
NUSE_PIPE pipe,
U8 *message,
U8 suspend);
```

Parameters:

pipe—the index (ID) of the pipe to be utilized;

message—a pointer to the message to be sent, which is a sequence of bytes as long as the configured message size of the pipe;

suspend—specification for task suspend; may be NUSE_NO_SUSPEND or NUSE_SUSPEND.

Returns:

NUSE_SUCCESS—The call was completed successfully.

NUSE_INVALID_PIPE—The pipe index is invalid.

NUSE_INVALID_POINTER—The message pointer is NULL.

NUSE_INVALID_SUSPEND—Suspend was attempted from a non-task thread or when blocking API calls were not enabled.

NUSE_PIPE_FULL—The pipe is full and suspend was not specified.

NUSE_PIPE_WAS_RESET—The pipe was reset, while the task was suspended.

NUSE_Queue_Count()

Obtains the number of queues configured in the application (see Chapter 12: *Queues*).

Service call prototype:

```
U8 NUSE_Queue_Count(void);
```

Parameters:

None

Returns:

The number of configured queues in the application

NUSE_Queue_Information()

Obtains information about a specific queue.

Service call prototype:

```
STATUS NUSE_Queue_Information(
NUSE_QUEUE queue,
ADDR *start_address,
U8 *queue_size,
U8 *available,
U8 *messages,
U8 *tasks_waiting,
NUSE_TASK *first_task);
```

Parameters:

queue—the index of the queue about which information is being requested;

start_address—a pointer to a variable of type ADDR, which will receive the address of the start of the queue's data area;

queue_size—a pointer to a variable of type U8, which will receive the total number of messages for which the queue has capacity;

available—a pointer to a variable of type U8, which will receive the number of messages for which the queue has currently remaining capacity;

messages—a pointer to a variable of type U8, which will receive the number of messages currently in the queue;

tasks_waiting—a pointer to a variable, which will receive the number of tasks suspended on this queue (nothing returned if task suspend is disabled);

first_task—a pointer to a variable of type NUSE_TASK, which will receive the index of the first suspended task (nothing returned if task suspend is disabled).

Returns:

NUSE_SUCCESS—The call was completed successfully.

NUSE_INVALID_QUEUE—The queue index is not valid.

NUSE_INVALID_POINTER—One or more of the pointer parameters is invalid.

NUSE_Queue_Jam()

Writes data to the front of a specified pipe.

Service call prototype:

```
STATUS NUSE_Queue_Jam(
NUSE_QUEUE queue,
ADDR *message,
U8 suspend);
```

Parameters:

queue—the index (ID) of the queue to be utilized;

message—a pointer to the message to be sent, which is a single variable of type ADDR;

suspend—specification for task suspend; may be NUSE_NO_SUSPEND or NUSE_SUSPEND.

Returns:

NUSE_SUCCESS—The call was completed successfully.

NUSE_INVALID_QUEUE—The queue index is invalid.

NUSE_INVALID_POINTER—The message pointer is NULL.

NUSE_INVALID_SUSPEND—Suspend was attempted from a non-task thread or when blocking API calls were not enabled.

NUSE_QUEUE_FULL—The queue is full and suspend was not specified.

NUSE_QUEUE_WAS_RESET—The queue was reset, while the task was suspended.

NUSE_Queue_Receive()

Reads data from a specified queue.

Service call prototype:

```
STATUS NUSE_Queue_Receive(
NUSE_QUEUE queue,
ADDR *message,
U8 suspend);
```

Parameters:

queue—the index (ID) of the queue to be utilized;

`message`—a pointer to storage for the message to be received, which is a single variable of type `ADDR`;

`suspend`—specification for task suspend; may be `NUSE_NO_SUSPEND` or `NUSE_SUSPEND`.

Returns:

`NUSE_SUCCESS`—The call was completed successfully.

`NUSE_INVALID_QUEUE`—The queue index is invalid.

`NUSE_INVALID_POINTER`—The message pointer is `NULL`.

`NUSE_INVALID_SUSPEND`—Suspend was attempted from a non-task thread or when blocking API calls were not enabled.

`NUSE_QUEUE_EMPTY`—The queue is empty and suspend was not specified.

`NUSE_QUEUE_WAS_RESET`—The queue was reset, while the task was suspended.

NUSE_Queue_Reset()

Restores a queue to its initial, unused state.

Service call prototype:

```
STATUS NUSE_Queue_Reset(NUSE_QUEUE queue);
```

Parameters:

`queue`—the index (ID) of the queue to be reset.

Returns:

`NUSE_SUCCESS`—The call was completed successfully.

`NUSE_INVALID_QUEUE`—The queue index is not valid.

NUSE_Queue_Send()

Writes data to the back of a specified queue.

Service call prototype:

```
STATUS NUSE_Queue_Send(
NUSE_QUEUE queue,
ADDR *message,
U8 suspend);
```

Parameters:

`queue`—the index (ID) of the queue to be utilized;

message—a pointer to the message to be sent, which is a single variable of type ADDR;

suspend—specification for task suspend; may be NUSE_NO_SUSPEND or NUSE_SUSPEND.

Returns:

NUSE_SUCCESS—The call was completed successfully.

NUSE_INVALID_QUEUE—The queue index is invalid.

NUSE_INVALID_POINTER—The message pointer is NULL.

NUSE_INVALID_SUSPEND—Suspend was attempted from a non-task thread or when blocking API calls were not enabled.

NUSE_QUEUE_FULL—The queue is full and suspend was not specified.

NUSE_QUEUE_WAS_RESET—The queue was reset, while the task was suspended.

NUSE_Release_Information()

Returns a pointer to a version string (see Chapter 18: *Diagnostics and Error Checking*).

Service call prototype:

```
char *NUSE_Release_Information(void);
```

Parameters:

None

Returns:

Pointer to NULL-terminated version string

NUSE_Semaphore_Count()

Obtains the number of semaphores configured in the application (see Chapter 10: *Semaphores*).

Service call prototype:

```
U8 NUSE_Semaphore_Count(void);
```

Parameters:

None

Returns:

The number of configured semaphores in the application

NUSE_Semaphore_Information()

Obtains information about a specific semaphore.

Service call prototype:

```
STATUS NUSE_Semaphore_Information(
NUSE_SEMAPHORE semaphore,
U8 *current_count,
U8 *tasks_waiting,
NUSE_TASK *first_task);
```

Parameters:

semaphore—the index of the semaphore about which information is being requested;

current_count—a pointer to a variable, which will receive the current value of the semaphore counter

tasks_waiting—a pointer to a variable, which will receive the number of tasks suspended on this semaphore (nothing returned if task suspend is disabled);

first_task—a pointer to a variable of type NUSE_TASK, which will receive the index of the first suspended task (nothing returned if task suspend is disabled).

Returns:

NUSE_SUCCESS—The call was completed successfully.

NUSE_INVALID_SEMAPHORE—The semaphore index is not valid.

NUSE_INVALID_POINTER—One or more of the pointer parameters is invalid.

NUSE_Sempahore_Obtain()

Obtains (decrements) a specified semaphore.

Service call prototype:

```
STATUS NUSE_Semaphore_Obtain(
NUSE_SEMAPHORE semaphore,
U8 suspend);
```

Parameters:

semaphore—the index (ID) of the semaphore to be utilized;

suspend—specification for task suspend; may be NUSE_NO_SUSPEND or NUSE_SUSPEND.

Returns:

NUSE_SUCCESS—The call was completed successfully.

NUSE_UNAVAILABLE—The semaphore had the value zero.

NUSE_INVALID_SEMAPHORE—The semaphore index is invalid.

NUSE_INVALID_SUSPEND—Suspend was attempted from a non-task thread or when blocking API calls were not enabled.

NUSE_SEMAPHORE_WAS_RESET—The semaphore was reset, while the task was suspended.

NUSE_Semaphore_Release()

Releases (increments) a specified semaphore.

Service call prototype:

STATUS NUSE_Semaphore_Release(NUSE_SEMAPHORE semaphore);

Parameters:

semaphore—the index (ID) of the semaphore to be released.

Returns:

NUSE_SUCCESS—The call was completed successfully.

NUSE_INVALID_SEMAPHORE—The semaphore index is invalid.

NUSE_UNAVAILABLE—The semaphore has the value 255 and cannot be incremented.

NUSE_Semaphore_Reset()

Restores a semaphore to its initial, unused state.

Service call prototype:

```
STATUS NUSE_Semaphore_Reset(
NUSE_SEMAPHORE semaphore,
U8 initial_count);
```

Parameters:

semaphore—the index (ID) of the semaphore to be reset;

initial_count—the vale to which the semaphore's counter is to be set.

Returns:

NUSE_SUCCESS—The call was completed successfully.

NUSE_INVALID_SEMAPHORE—The semaphore index is not valid.

NUSE_Signals_Receive()

Destructively reads and returns the current task's signal flags (see Chapter 8: *Signals*).

Service call prototype:

```
U8 NUSE_Signals_Receive(void);
```

Parameters:

None

Returns:

The signals' flags value

NUSE_Signals_Send()

Sets selected signal flags for a specified task.

Service call prototype:

```
STATUS NUSE_Signals_Send(
NUSE_TASK task,
U8 signals);
```

Parameters:

task—the index (ID) of the task that owns the signal flags to be set;

signals—the value of the signal flags to be set.

Returns:

NUSE_SUCCESS—The call was completed successfully.

NUSE_INVALID_TASK—The task index is invalid.

NUSE_Task_Check_Stack()

Returns the approximate remaining stack space for the current task (see Chapter 6: *Tasks*).

Service call prototype:

```
U16 NUSE_Task_Check_Stack(U8 dummy);
```

Parameters:

dummy—any value may be provided, as it is not actually used.

Returns:

Number of bytes of available stack for the current task.

NUSE_Task_Count()

Obtains the number of tasks configured in the application.

Service call prototype:

```
U8 NUSE_Task_Count(void);
```

Parameters:

None

Returns:

The number of configured tasks in the application

NUSE_Task_Current()

Returns the ID (index) of the current task.

Service call prototype:

```
NUSE_TASK NUSE_Task_Current(void);
```

Parameters:

None

Returns:

Index of the current (calling) task

NUSE_Task_Information()

Obtains information about a specific task.

Service call prototype:

```
STATUS NUSE_Task_Information(
NUSE_TASK task,
U8 *task_status,
U16 *scheduled_count,
ADDR *stack_base,
U16 *stack_size);
```

Parameters:

task—the index of the task about which information is being requested;

`task_status`—a pointer to a `U8` variable, which will receive the current value of the task status (nothing returned if task suspend is disabled);

`scheduled_count`—a pointer to a `U16` variable, which will receive a count of the number of times this task has been scheduled (nothing returned if schedule count is disabled);

`stack_base`—a pointer to an `ADDR` variable, which will receive the address of the task's stack (nothing returned if the RTC scheduler is in use);

`stack_size`—a pointer to a `U16` variable, which will receive the size of the task's stack (nothing returned if the RTC scheduler is in use).

Returns:

`NUSE_SUCCESS`—The call was completed successfully.

`NUSE_INVALID_TASK`—The task index is not valid.

`NUSE_INVALID_POINTER`—One or more of the pointer parameters is invalid.

NUSE_Task_Relinquish()

Relinquishes control of the CPU to the next available task (when using round robin or time-sliced scheduler).

Service call prototype:

`void NUSE_Task_Relinquish(void);`

Parameters:

None

Returns:

Nothing

NUSE_Task_Reset()

Restores a task to its initial, unused state and places it in an unconditional suspend state.

Service call prototype:

`STATUS NUSE_Task_Reset(NUSE_TASK task);`

Parameters:

`task`—the index (ID) of the task to be reset.

Returns:

> NUSE_SUCCESS—The call was completed successfully.

> NUSE_INVALID_TASK—The task index is not valid.

NUSE_Task_Resume()

Wakes up a specified task (which was unconditionally suspended).

Service call prototype:

> STATUS NUSE_Task_Resume(NUSE_TASK task);

Parameters:

> task—the index (ID) of the task to be resumed.

Returns:

> NUSE_SUCCESS—The call was completed successfully.

> NUSE_INVALID_TASK—The task index is invalid.

> NUSE_INVALID_RESUME—The task was not unconditionally suspended.

NUSE_Task_Sleep()

Places the current task in sleep suspend for a specified period.

Service call prototype:

> void NUSE_Task_Sleep(U16 ticks);

Parameters:

> ticks—the time period, in real-time clock ticks, for which the task should be suspended.

Returns:

> Nothing.

NUSE_Task_Suspend()

Places a specified task into unconditional suspend.

Service call prototype:

> STATUS NUSE_Task_Suspend(NUSE_TASK task);

Parameters:

> task—the index (ID) of the task to be suspended (which may be the current task, the ID of which may be obtained using NUSE_Task_Current()).

Returns:

> **NUSE_SUCCESS**—The call was completed successfully.
>
> **NUSE_INVALID_TASK**—The task index is invalid.

NUSE_Timer_Control()

Enables/disables (starts/stops) a specified application timer (see Chapter 15: *Application Timers*).

Service call prototype:

```
STATUS NUSE_Timer_Control(
NUSE_TIMER timer,
OPTION enable);
```

Parameters:

> **timer**—the index (ID) of the timer to be utilized;
>
> **enable**—required function; may be **NUSE_ENABLE_TIMER** or **NUSE_DISABLE_TIMER**.

Returns:

> **NUSE_SUCCESS**—The call was completed successfully.
>
> **NUSE_INVALID_TIMER**—The timer index is invalid.
>
> **NUSE_INVALID_ENABLE**—The specified function is invalid.

NUSE_Timer_Count()

Obtains the number of application timers configured in the application.

Service call prototype:

```
U8 NUSE_Timer_Count(void);
```

Parameters:

None

Returns:

The number of configured timers in the application

NUSE_Timer_Get_Remaining()

Returns the remaining time (in ticks) before a specified application timer expires.

Service call prototype:

```
STATUS NUSE_Timer_Get_Remaining(
NUSE_TIMER timer,
U16 *remaining_time);
```

Parameters:

timer—the index (ID) of the timer to be utilized;

remaining_time—a pointer to storage for the remaining time value, which is a single variable of type U16.

Returns:

NUSE_SUCCESS—The call was completed successfully.

NUSE_INVALID_TIMER—The timer index is invalid.

NUSE_INVALID_POINTER—The remaining time pointer is NULL.

NUSE_Timer_Information()

Obtains information about a specific application timer.

Service call prototype:

```
STATUS NUSE_Timer_Information(
NUSE_TIMER timer,
OPTION *enable,
U8 *expirations,
U8 *id,
U16 *initial_time,
U16 *reschedule_time);
```

Parameters:

timer—the index of the timer about which information is being requested;

enable—a pointer to a variable, which will receive a value of TRUE or FALSE depending on whether the timer is enabled or not;

expirations—a pointer to a variable of type U8, which will receive the count of the number of times the timer has expired since it was last reset;

initial_time—a pointer to a variable of type U16, which will receive the value to which the timer is initialized on reset;

reschedule_time—a pointer to a variable of type U16, which will receive the value to which the timer is initialized on expiration;

id—a pointer to a variable of type u8, which will receive the value of the parameter passed to the timer's expiration routine (nothing returned if expiration routines are disabled).

Returns:

NUSE_SUCCESS—The call was completed successfully.

NUSE_INVALID_TIMER—The timer index is not valid.

NUSE_INVALID_POINTER—One or more of the pointer parameters is invalid.

NUSE_Timer_Reset()

Restores an application timer to its initial, unused state.

Service call prototype:

```
STATUS NUSE_Timer_Reset(
NUSE_TIMER timer,
OPTION enable);
```

Parameters:

timer—the index (ID) of the timer to be reset;

enable—required state after reset; may be NUSE_ENABLE_TIMER or NUSE_DISABLE_TIMER.

Returns:

NUSE_SUCCESS—The call was completed successfully.

NUSE_INVALID_TIMER—The timer index is not valid.

NUSE_INVALID_ENABLE—The specified state is invalid.

Nucleus SE symbols and data structures

Throughout this book, all the symbols and data structures used by Nucleus SE are documented in context. Here, they are all brought together for quick reference, categorized and presented in alphabetical order.

Naming conventions

It is good programming practice to maintain a naming convention for all identifiers and this approach is used by all the Nucleus SE code. This is particularly important, as all the

Nucleus SE symbols are "visible" to the applications code and naming clashes would be a possibility, leading to bugs and confusion.

Almost all identifiers used by Nucleus SE code have a prefix NUSE_ (which is an abbreviation for NUCLEUS SE). There are a few exceptions, which are outlined later in this chapter.

All variables have names using mixed case and take the form NUSE_Category_Name, which is intended to give some indication of its usage; an example is: NUSE_Task_State.

All constants have uppercase names, mostly employing the NUSE_ prefix; an example is: NUSE_QUEUE_EMPTY.

Function names are mixed case, and all carry the NUSE_ prefix. API function names all have the form NUSE_Category_Name(); an example is: NUSE_Partition_Allocate(). Other functions, which are not normally used by application code, have the form NUSE_Name(); an example is: NUSE_Scheduler().

Constants—#define symbols and macros

The Nucleus SE code makes extensive use of #define symbols and a few macros. These are intended to make code easier to read and simplify migration to Nucleus RTOS. All the symbols are defined in a few header files, which may be easily included in an application module by just including nuse.h.

API functions

A number of symbols are employed to represent parameters to API functions and return/error/status codes. They are all defined in nuse_codes.h.

Parameter values—A number of API functions have parameters that require an arbitrary value to specify a requirement. The following symbols represent these values:

NUSE_AND—used in calls to NUSE_Event_Group_Set() and NUSE_Event_Group_Retrieve() to indicate the logic of event flag operations.

NUSE_AND_CONSUME—unused; included for compatibility and possible future application.

NUSE_DISABLE_TIMER—parameter value used in calls to NUSE_Timer_Control() to disable a timer.

NUSE_ENABLE_TIMER—parameter value used in calls to NUSE_Timer_Control() to enable a timer.

`NUSE_FALSE`—unused; included for compatibility and possible future application.

`NUSE_FIFO`—unused; included for compatibility and possible future application.

`NUSE_FIXED_SIZE`—unused; included for compatibility and possible future application.

`NUSE_NO_SUSPEND`—used in API function calls, where blocking is permitted, to indicate that the function should return immediately and not block on resource availability.

`NUSE_NO_TASK`—a value used in calls to `NUSE_Reschedule()`, which indicates that the caller cannot give a hint about which task may be scheduled next. This is not recommended for use by application code.

`NUSE_OR`—used in calls to `NUSE_Event_Group_Set()` and `NUSE_Event_Group_Retrieve()` to indicate the logic of event flag operations.

`NUSE_OR_CONSUME`—unused; included for compatibility and possible future application.

`NUSE_PRIORITY`—unused; included for compatibility and possible future application.

`NUSE_SUSPEND`—used in API function calls, where blocking is permitted, to indicate that the function should not return immediately and block pending resource availability.

`NUSE_TRUE`—unused; included for compatibility and possible future application.

`NUSE_VARIABLE_SIZE`—unused; included for compatibility and possible future application.

Return/error/status codes—Many API functions return results as a function return value. Frequently, these results are arbitrary values that may be represented by symbols:

`NUSE_INVALID_ENABLE`—indicates that the enable specifying parameter in a call to `NUSE_Timer_Control()` is invalid (i.e., not `NUSE_ENABLE_TIMER` or `NUSE_DISABLE_TIMER`).

`NUSE_INVALID_FUNCTION`—unused; included for compatibility and possible future application.

`NUSE_INVALID_GROUP`—a return value which indicates that the event group index, provided in a call to

NUSE_Event_Group_Set(), NUSE_Event_Group_Retrieve(), or NUSE_Event_Group_Information(), is invalid (i.e., too large a value).

NUSE_INVALID_MAILBOX—indicates that the mailbox index, provided in a call to NUSE_Mailbox_Send(), NUSE_Mailbox_Receive(), NUSE_Mailbox_Reset(), or NUSE_Mailbox_Information(), is invalid (i.e., too large a value).

NUSE_INVALID_OPERATION—indicates that the operation specifying parameter in a call to NUSE_Event_Group_Set() or NUSE_Event_Group_Retrieve() is invalid (i.e., not NUSE_OR or NUSE_AND).

NUSE_INVALID_PIPE—indicates that the pipe index, provided in a call to NUSE_Pipe_Send(), NUSE_Pipe_Receive(), NUSE_Pipe_Jam(), NUSE_Pipe_Reset(), or NUSE_Pipe_Information(), is invalid (i.e., too large a value).

NUSE_INVALID_POINTER—indicates that an API function has been called, which requires one or more pointer parameters (to return data), and one of those pointers is NULL.

NUSE_INVALID_POOL—indicates that the partition pool index, provided in a call to NUSE_Partition_Allocate() or NUSE_Partition_Pool_Information(), is invalid (i.e., too large a value).

NUSE_INVALID_QUEUE—indicates that the queue index, provided in a call to NUSE_Queue_Send(), NUSE_Queue_Receive(), NUSE_Queue_Jam(), NUSE_Queue_Reset(), or NUSE_Queue_Information(), is invalid (i.e., too large a value).

NUSE_INVALID_RESUME—indicates that the task specified in a call to NUSE_Task_Resume() is not unconditionally suspended, which is a requirement for it to be resumed.

NUSE_INVALID_SEMAPHORE—a return value which indicates that the semaphore index, provided in a call to NUSE_Semaphore_Obtain(), NUSE_Semaphore_Release(), NUSE_Semaphore_Reset(), or NUSE_Semaphore_Information(), is invalid (i.e., too large a value).

NUSE_INVALID_SIZE—indicates that the message size, provided in a call to NUSE_Pipe_Send(), NUSE_Pipe_Receive(), or NUSE_Pipe_Jam() does not correspond to that configured for the pipe.

NUSE_INVALID_SUSPEND—indicates that an API function has received an invalid suspend request; this may be because it has an invalid value (i.e., something other than `NUSE_SUSPEND` or `NUSE_NO_SUSPEND`) or `NUSE_SUSPEND` is specified, but task blocking is not enabled.

`NUSE_INVALID_TASK`—indicates that the task index, provided in a call to `NUSE_Signals_Send()`, `NUSE_Task_Suspend()`, `NUSE_Task_Resume()`, `NUSE_Task_Reset()`, or `NUSE_Task_Information()`, is invalid (i.e., too large a value).

`NUSE_INVALID_TIMER`—indicates that the timer index, provided in a call to `NUSE_Timer_Control()`, `NUSE_Timer_Get_Remaining()`, `NUSE_Timer_Reset()`, or `NUSE_Timer_Information()`, is invalid (i.e., too large a value).

`NUSE_MAILBOX_EMPTY`—indicates that a call to `NUSE_Mailbox_Receive()`, with a `NUSE_NO_SUSPEND` parameter, was made and the specified mailbox did not contain a message.

`NUSE_MAILBOX_FULL`—indicates that a call to `NUSE_Mailbox_Send()`, with a `NUSE_NO_SUSPEND` parameter, was made and the specified mailbox already contained a message.

`NUSE_MAILBOX_WAS_RESET`—a return value from a call to `NUSE_Mailbox_Send()` or `NUSE_Mailbox_Receive()`, with a `NUSE_SUSPEND` parameter, which resulted in the calling task being suspended. The specified mailbox was reset by another task making a call to `NUSE_Mailbox_Reset()`, which resulted in the calling task being resumed and this value being returned.

`NUSE_NO_PARTITION`—indicates that a call to `NUSE_Partition_Allocate()`, with a `NUSE_NO_SUSPEND` parameter, was made and there were no free partitions in the specified partition pool.

`NUSE_NOT_DISABLED`—indicates that a call to `NUSE_Timer_Reset()` was made, but the specified timer was still enabled.

`NUSE_NOT_PRESENT`—indicates that a call to `NUSE_Event_Group_Retrieve()` was made, with a `NUSE_NO_SUSPEND` parameter, but no event flags were retrieved (none present for an or operation; not all present for and AND).

`NUSE_NOT_TERMINATED`—unused; included for compatibility and possible future application.

`NUSE_PIPE_EMPTY`—indicates that a call to `NUSE_Pipe_Receive()`, with a `NUSE_NO_SUSPEND` parameter, was made and there were no messages in the specified pipe.

NUSE_PIPE_FULL—indicates that a call to NUSE_Pipe_Send() or NUSE_Pipe_Jam(), with a NUSE_NO_SUSPEND parameter, was made and there were no message slots available in the specified pipe.

NUSE_PIPE_WAS_RESET—a return value from a call to NUSE_Pipe_Send(), NUSE_Pipe_Jam(), or NUSE_Pipe_Receive(), with a NUSE_SUSPEND parameter, which resulted in the calling task being suspended. The specified pipe was reset by another task making a call to NUSE_Pipe_Reset(), which resulted in the calling task being resumed and this value being returned.

NUSE_QUEUE_EMPTY—indicates that a call to NUSE_Queue_Receive(), with a NUSE_NO_SUSPEND parameter, was made and there were no messages in the specified queue.

NUSE_QUEUE_FULL—indicates that a call to NUSE_Queue_Send() or NUSE_Queue_Jam(), with a NUSE_NO_SUSPEND parameter, was made and there were no message slots available in the specified queue.

NUSE_QUEUE_WAS_RESET—a return value from a call to NUSE_Queue_Send(), NUSE_Queue_Jam(), or NUSE_Queue_Receive(), with a NUSE_SUSPEND parameter, which resulted in the calling task being suspended. The specified queue was reset by another task making a call to NUSE_Queue_Reset(), which resulted in the calling task being resumed and this value being returned.

NUSE_SEMAPHORE_WAS_RESET—a return value from a call to NUSE_Semaphore_Obtain(), with a NUSE_SUSPEND parameter, which resulted in the calling task being suspended. The specified semaphore was reset by another task making a call to NUSE_Semaphore_Reset(), which resulted in the calling task being resumed and this value being returned.

NUSE_SUCCESS—indicates that an API function has completed successfully.

NUSE_TIMEOUT—unused; included for compatibility and possible future application.

NUSE_UNAVAILABLE—a return value from a call to NUSE_Semaphore_Obtain() where the semaphore had the value 0 and could not, therefore, be obtained (decremented) or a call to NUSE_Semaphore_Release() where the semaphore had the value 255 and could not, therefore, be released (incremented).

Status values

A number of symbols are employed to indicate the state of the executing code or the status of a task. They are all defined in `nuse_codes.h`.

Thread state—A value in the global variable `NUSE_Task_State` (see "*Global data*" section later in this chapter) indicates the state of currently executing code:

> `NUSE_MISR_CONTEXT`—The current code is running in the context of a managed interrupt service routine.

> `NUSE_NISR_CONTEXT`—The current code is running in the context of a native interrupt service routine.

> `NUSE_STARTUP_CONTEXT`—The current code is running in the context of the start-up/initialization sequence that is executed prior to the scheduler being started.

> `NUSE_TASK_CONTEXT`—The current code is running in the context of an application task.

Task status—If task suspend is enabled in Nucleus SE, an array, `NUSE_Task_Status[]` (see "*Kernel object tables*" section later in this chapter), contains values to indicate the status of a task with respect to its availability for scheduling. These values are stored in the lower nibble of the array element; the upper nibble commonly holds the index of object on which a task is suspended:

> `NUSE_EVENT_SUSPEND`—Task is suspended on an event flag group as a result of a call to `NUSE_Event_Group_Retrieve()` with a `NUSE_SUSPEND` parameter.

> `NUSE_FINISHED`—unused; included for compatibility and possible future application.

> `NUSE_MAILBOX_SUSPEND`—Task is suspended on a mailbox as a result of a call to `NUSE_Mailbox_Send()` or `NUSE_Mailbox_Receive()` with a `NUSE_SUSPEND` parameter.

> `NUSE_PARTITION_SUSPEND`—Task is suspended on a partition pool as a result of a call to `NUSE_Partition_Allocate()` with a `NUSE_SUSPEND` parameter.

> `NUSE_PIPE_SUSPEND`—Task is suspended on a pipe as a result of a call to `NUSE_Pipe_Send()`, `NUSE_Pipe_Jam()`, or `NUSE_Pipe_Receive()` with a `NUSE_SUSPEND` parameter.

> `NUSE_PURE_SUSPEND`—Task is unconditionally suspended as a result of a call to `NUSE_Task_Suspend()` or `NUSE_Task_Reset()`.

NUSE_QUEUE_SUSPEND—Task is suspended on a queue as a result of a call to **NUSE_Queue_Send()**, **NUSE_Queue_Jam()**, or **NUSE_Queue_Receive()** with a **NUSE_SUSPEND** parameter.

NUSE_READY—Task is not suspended and is available to be scheduled.

NUSE_SEMAPHORE_SUSPEND—Task is suspended on a semaphore as a result of a call to **NUSE_Semaphore_Obtain()** with a **NUSE_SUSPEND** parameter.

NUSE_SLEEP_SUSPEND—Task is suspended as a result of a call to **NUSE_Task_Sleep()**.

NUSE_TERMINATED—unused; included for compatibility and possible future application.

Miscellaneous symbols

A number of symbols are employed to simplify porting of Nucleus SE to different toolkits and processors. They are all defined in **nuse_types.h**:

FALSE—logical false value, which is normally set to **(0)**.

HINIB()—macro, which extracts the high nibble of a byte.

INTERRUPT—symbol used to qualify a function definition to indicate that it is to be compiled as an interrupt service routine; commonly set to the extension keyword **interrupt**.

LONIB()—macro, which extracts the low nibble of a byte.

NULL—symbol used to signify a null pointer; normally set to **(0)**.

NUSE_CONTEXT_SWAP()—a macro, which is used by the scheduler to invoke a task context swap; this is normally an assembly language insert that facilitates a "trap" or "software interrupt" so that a status register is pushed onto the stack along with the return address; this is totally CPU-specific.

NUSE_CS_Enter()—a macro, used by many API functions to delineate the start of a "critical section"; that is, a sequence of noninterruptible instructions; this may be an assembly language insert to disable interrupts, for example; this is totally CPU-specific.

NUSE_CS_Exit()—a macro, used by many API functions to delineate the end of a "critical section"; that is, a sequence of noninterruptible instructions; this may be an assembly language insert to re-enable interrupts, for example; this is totally CPU-specific.

NUSE_MANAGED_ISR()—a macro used as a "wrapper" to create a managed interrupt service routine; call to the provided function is surrounded by a pair of assembly language inserts that facilitate task context save and restore; this is totally CPU-specific.

NUSE_NISR_Enter()—a macro used to indicate the start of a native interrupt service routine; it simply saves the current value of the thread state (**NUSE_Task_State**) and loads a new value of **NUSE_NISR_CONTEXT**.

NUSE_NISR_Exit()—a macro used to indicate the end of a native interrupt service routine; it restores the previous value of the thread state (**NUSE_Task_State**).

NUSE_REGISTERS—a symbol that specifies how many CPU registers are to be included in a context save; it is used to size the relevant data structures; this is totally CPU-specific.

NUSE_STATUS_REGISTER—a symbol that represents the "normal" value of the processor's status register; this is totally CPU-specific.

RAM—a symbol used to qualify a variable definition to signify that it is to be located in read/write memory; normally left undefined.

ROM—a symbol used to qualify a variable definition to signify that it is to be located in read-only memory; normally set to **const**.

TRUE—logical true value, which is normally set to **(1)**.

System configuration

A Nucleus SE application is largely configured by setting values of symbols in **nuse_config.h,** which select the number of each object type, where API functions are enabled, the scheduler configuration, and some other optional facilities.

Tasks—The number of tasks in an application is determined by the setting of **NUSE_TASK_NUMBER,** which can take a value in the range 1–16.

Each API function may be enabled or disabled by setting its control symbol to **TRUE** or **FALSE,** respectively. The symbols have the same name as the API function, as follows:

```
NUSE_TASK_CHECK_STACK
NUSE_TASK_COUNT
NUSE_TASK_CURRENT
NUSE_TASK_INFORMATION
NUSE_TASK_RELINQUISH
NUSE_TASK_RESET
NUSE_TASK_RESUME
NUSE_TASK_SLEEP
NUSE_TASK_SUSPEND
```

Additionally, the symbol `NUSE_INITIAL_TASK_STATE_SUPPORT` enables a facility that permits the start-up status of a task to be specified (as `NUSE_READY` or `NUSE_PURE_SUSPEND`); otherwise all tasks are ready to run.

Partition pools—The number of partition pools in an application is determined by the setting of `NUSE_PARTITION_POOL_NUMBER`, which can take a value in the range 0–16.

Each API function may be enabled or disabled by setting its control symbol to `TRUE` or `FALSE`, respectively. The symbols have the same name as the API function, as follows:

```
NUSE_PARTITION_ALLOCATE
NUSE_PARTITION_DEALLOCATE
NUSE_PARTITION_POOL_COUNT
NUSE_PARTITION_POOL_INFORMATION
```

Mailboxes—The number of mailboxes in an application is determined by the setting of `NUSE_MAILBOX_NUMBER`, which can take a value in the range 0–16.

Each API function may be enabled or disabled by setting its control symbol to `TRUE` or `FALSE`, respectively. The symbols have the same name as the API function, as follows:

```
NUSE_MAILBOX_COUNT
NUSE_MAILBOX_INFORMATION
NUSE_MAILBOX_RECEIVE
NUSE_MAILBOX_RESET
NUSE_MAILBOX_SEND
```

Queues—The number of queues in an application is determined by the setting of `NUSE_QUEUE_NUMBER`, which can take a value in the range 0–16.

Each API function may be enabled or disabled by setting its control symbol to `TRUE` or `FALSE`, respectively. The symbols have the same name as the API function, as follows:

```
NUSE_QUEUE_COUNT
NUSE_QUEUE_INFORMATION
NUSE_QUEUE_JAM
```

```
NUSE_QUEUE_RECEIVE
NUSE_QUEUE_RESET
NUSE_QUEUE_SEND
```

Pipes—The number of pipes in an application is determined by the setting of `NUSE_PIPE_NUMBER`, which can take a value in the range 0–16.

Each API function may be enabled or disabled by setting its control symbol to `TRUE` or `FALSE`, respectively. The symbols have the same name as the API function, as follows:

```
NUSE_PIPE_COUNT
NUSE_PIPE_INFORMATION
NUSE_PIPE_JAM
NUSE_PIPE_RECEIVE
NUSE_PIPE_RESET
NUSE_PIPE_SEND
```

Semaphores—The number of semaphores in an application is determined by the setting of `NUSE_SEMAPHORE_NUMBER`, which can take a value in the range 0–16.

Each API function may be enabled or disabled by setting its control symbol to `TRUE` or `FALSE`, respectively. The symbols have the same name as the API function, as follows:

```
NUSE_SEMAPHORE_COUNT
NUSE_SEMAPHORE_INFORMATION
NUSE_SEMAPHORE_OBTAIN
NUSE_SEMAPHORE_RELEASE
NUSE_SEMAPHORE_RESET
```

Event groups—The number of event groups in an application is determined by the setting of `NUSE_EVENT_GROUP_NUMBER`, which can take a value in the range 0–16.

Each API function may be enabled or disabled by setting its control symbol to `TRUE` or `FALSE`, respectively. The symbols have the same name as the API function, as follows:

```
NUSE_EVENT_GROUP_COUNT
NUSE_EVENT_GROUP_INFORMATION
NUSE_EVENT_GROUP_RETRIEVE
NUSE_EVENT_GROUP_SET
```

Signals—As signals are not independent objects, but are, rather, associated with tasks, there is not a question of specifying their number. If they are enabled, by setting the symbol `NUSE_SIGNAL_SUPPORT` to `TRUE`, a set of eight signals is provided for each task in the application.

Each API function may be enabled or disabled by setting its control symbol to TRUE or FALSE, respectively. The symbols have the same name as the API function, as follows:

```
NUSE_SIGNALS_RECEIVE
NUSE_SIGNALS_SEND
```

Application timers—The number of timers in an application is determined by the setting of NUSE_TIMER_NUMBER, which can take a value in the range 0–16.

Each API function may be enabled or disabled by setting its control symbol to TRUE or FALSE, respectively. The symbols have the same name as the API function, as follows:

```
NUSE_TIMER_CONTROL
NUSE_TIMER_COUNT
NUSE_TIMER_GET_REMAINING
NUSE_TIMER_INFORMATION
NUSE_TIMER_RESET
```

Additionally, the symbol NUSE_TIMER_EXPIRATION_ROUTINE_SUPPORT enables support for timer expiration routines, which is optional in Nucleus SE.

System time—The tick clock is enabled using the symbol NUSE_SYSTEM_TIME_SUPPORT.

Each API function may be enabled or disabled by setting its control symbol to TRUE or FALSE, respectively. The symbols have the same name as the API function, as follows:

```
NUSE_CLOCK_RETRIEVE
NUSE_CLOCK_SET
```

Diagnostics—The optional checking of API function parameters may be enabled by use of the symbol NUSE_API_PARAMETER_CHECKING.

The only API function that provides diagnostic facilities may be enabled via NUSE_RELEASE_INFORMATION.

Scheduler—Nucleus SE has a choice of schedulers. These are selected by setting the symbol NUSE_SCHEDULER_TYPE to one of:

```
NUSE_RUN_TO_COMPLETION_SCHEDULER
NUSE_TIME_SLICE_SCHEDULER
NUSE_ROUND_ROBIN_SCHEDULER
NUSE_PRIORITY_SCHEDULER
```

If the time slice scheduler is selected, the number of ticks per slice may be selected by assigning a value to NUSE_TIME_SLICE_TICKS.

A count may be maintained of how many times each task is scheduled. This facility is enabled by means of the symbol `NUSE_SCHEDULE_COUNT_SUPPORT`.

Although the ability to suspend tasks is useful, this capability is optional in Nucleus SE (except when the priority scheduler is in use) and is enabled via `NUSE_SUSPEND_ENABLE`.

The optional facility to perform API function calls that suspend pending resource availability may be enabled with `NUSE_BLOCKING_ENABLE`.

Other functionality—An additional symbol—`NUSE_INCLUDE_EVERYTHING`—may be set when testing a new port of Nucleus SE, particularly when a new toolkit is employed. This option ensures that every possible option and all available API functions are enabled, thus providing the maximum amount of code for verification. It is simply a shortcut that avoids the need to set all the options by hand and should only be used for testing purposes.

Header file inclusion checks

Since Nucleus SE has a large number of header files, some care is needed to avoid multiple inclusions, which can lead to errors. A simple checking mechanism is employed. Every file includes a marker symbol, which is derived from the filename: dots are replaced by underscores and additional underscores are added to the start and end of the symbol; for example, the symbol for `nuse_codes.h` is `_NUSE_CODES_H_`. Conditional compilation ensures that the contents of the file are not included if the relevant symbol already exists.

Typedefs

In order to ensure code portability, the Nucleus SE code uses a defined set of data types, as the implementation of the C language intrinsic data types (`int`, `char`, `long`, etc.) is compiler-dependent. Although `#define` symbols could have been used, the C language `typedef` facility produces more robust code. These are defined in `nuse_types.h`.

Rational data types

The general-purpose data types are defined as follows:

    `U8`—unsigned 8 bit

    `S8`—signed 8 bit

U16—unsigned 16 bit

U32—unsigned 32 bit

ADDR—pointer/address

Kernel object indexes

To improve error checking and maintain some compatibility with the coding style of Nucleus RTOS, additional data types are defined for kernel object indexes:

```
NUSE_TASK
NUSE_PARTITION_POOL
NUSE_MAILBOX
NUSE_QUEUE
NUSE_PIPE
NUSE_SEMAPHORE
NUSE_EVENT_GROUP
NUSE_TIMER
```

These are all equated to U8.

Other data types

Some additional data types are also included:

OPTION—used for API function call parameters; normally U8.

PF0—pointer to a function with no parameters; used for syntactic clarity in the code dealing with task start-up.

PF1—pointer to a function with one U8 parameter; used for syntactic clarity in the code dealing with the calling of timer expiration routines.

PF2—pointer to a function with two parameters; unused and included for possible future applications.

STATUS—used for API function status return values.

Global data

Nucleus SE has a limited number of global variables, other than those that describe kernel objects, which are covered later in this appendix. The majority of the globals are held in read/write memory (RAM) and one is stored in read-only memory (ROM).

RAM

The RAM globals are all defined in `nuse_globals.c` and mostly initialized in `nuse_init.c`.

`NUSE_Task_Active`—index of the currently running task; type `NUSE_TASK`.

`NUSE_Task_Next`—index of the next task to be scheduled; type `NUSE_TASK`; only exists if a scheduler other than run to completion is selected.

`NUSE_Task_Saved_State`—previous value of `NUSE_Task_State` preserved during a managed interrupt service routine; type `U8`.

`NUSE_Task_State`—current thread state; type `U8`; may be `NUSE_MISR_CONTEXT`, `NUSE_NISR_CONTEXT`, `NUSE_STARTUP_CONTEXT`, or `NUSE_TASK_CONTEXT`.

`NUSE_Tick_Clock`—system time counter; type `U32`; this facility is optional.

`NUSE_Time_Slice_Ticks`—time slice tick down-counter; type `U16`; only exists when the time-sliced scheduler is in use.

ROM

There is only a single ROM-based global: `NUSE_Release_Info`. This is defined and initialized in `nuse_globals.c`. It is of type `char*` and points to a string of the form `"Nucleus SE - Xyymmdd"`, where `Xyymmdd` signifies the release type and date.

Kernel object tables

In Nucleus SE, all kernel objects are described and controlled by a number of tables (single-dimensional arrays) in RAM or ROM. With the single exception of `NUSE_Task_Start_Address[]`, all of these data structures are optional—their existence is determined by the system configuration and selected options.

All RAM-based data structures are defined and initialized in `nuse_init.c`.

All ROM-based data structures are defined and initialized (statically by the user) in `nuse_config.c`.

Tasks

Since there must (obviously) be at least one task in a Nucleus SE application, some of these data structures will always exist, depending on which options are selected in the configuration.

RAM

There are up to six RAM-based data structures appertaining to tasks:

NUSE_Task_Blocking_Return[]—used to hold the return code from an API function call where a task is suspended, which is typically NUSE_SUCCESS or an indication that the object was reset; type u8; only exists if task blocking is enabled.

NUSE_Task_Context[][]—used to hold the tasks' contexts (register sets), while they are not active; type ADDR, size is CPU-dependent; only exists if a scheduler other than run to completion is selected.

NUSE_Task_Schedule_Count[]—holds a count of the number of times a task has been made active since the system was initialized; type u16; only exists if schedule counting is enabled.

NUSE_Task_Signal_Flags[]—holds the set of eight signal flags for each task; type u8; only exists if signals are enabled.

NUSE_Task_Status[]—contains the current status of tasks: either NUSE_READY or an indication of why a task is suspended; type u8; only exists if task suspend is enabled.

NUSE_Task_Timeout_Counter[]—contains the down-counter that supports the NUSE_Task_Sleep() API function; type u16; only exists if this API function is enabled.

ROM

There are up to four ROM-based data structures appertaining to tasks:

NUSE_Task_Initial_State[]—contains the status for each task (NUSE_READY or NUSE_PURE_SUSPEND) when the scheduler is started; type u8; only exists if initial task state is enabled.

NUSE_Task_Stack_Base[]—contains the start address of the areas of memory allocated as stack space for each task; type **ADDR**; only exists if a scheduler other than run to completion is selected.

NUSE_Task_Stack_Size[]—contains the size (in words) of the areas of memory allocated as stack space for each task; type **U16**; only exists if a scheduler other than run to completion is selected.

NUSE_Task_Start_Address[]—contains the start address of the code for each task (i.e., a pointer to the outer function of each tasks' code); type **ADDR**; this data structure always exists.

Partition pools

Data structures appertaining to partition pools only exist if one or more partition pools are configured.

RAM

One or two RAM-based data structures may exist that support partition pools:

NUSE_Partition_Pool_Blocking_Count[]—count of how many tasks are suspended (blocked) on each pool pending availability of a partition; type **U8**; only exists if task blocking is enabled and at least one partition pool is configured.

NUSE_Partition_Pool_Partition_Used[]—count of how many partitions are currently in use in each pool; type **U8**; only exists if at least one partition pool is configured.

ROM

Three ROM-based data structures may exist that support partition pools:

NUSE_Partition_Pool_Data_Address[]—contains the start addresses of the memory areas allocated to each partition pool; type **ADDR**; only exists if at least one partition pool is configured.

NUSE_Partition_Pool_Partition_Number[]—contains the number of partitions in each pool; type **U8**; only exists if at least one partition pool is configured.

`NUSE_Partition_Pool_Partition_Size[]`—contains the size (in bytes) of the partitions in each pool; type `U16`; only exists if at least one partition pool is configured.

Mailboxes

Data structures appertaining to mailboxes only exist if one or more mailboxes are configured.

RAM

Two or three RAM-based data structures may exist that support mailboxes:

`NUSE_Mailbox_Blocking_Count[]`—count of how many tasks are suspended (blocked) on each mailbox pending availability of space to write data or the arrival of incoming data; type `U8`; only exists if task blocking is enabled and at least one mailbox is configured.

`NUSE_Mailbox_Data[]`—contains the data for each mailbox; type `ADDR`; only exists if at least one mailbox is configured.

`NUSE_Mailbox_Status[]`—contains status values (`TRUE` indicates full; `FALSE` indicate empty) for each mailbox; type `U8`; only exists if at least one mailbox is configured.

ROM

No ROM-based data structures exist that support mailboxes.

Queues

Data structures appertaining to queues only exist if one or more queues are configured.

RAM

Three or four RAM-based data structures may exist that support queues:

`NUSE_Queue_Blocking_Count[]`—count of how many tasks are suspended (blocked) on each queue pending availability of space to write data or the arrival of incoming data; type `U8`; only exists if task blocking is enabled and at least one queue is configured.

NUSE_Queue_Head[]—index of the head (location to which data will normally be written) for each queue; type **U8**; only exists if at least one queue is configured.

NUSE_Queue_Items[]—count of the number of data items stored in each queue; type **U8**; only exists if at least one queue is configured.

NUSE_Queue_Tail[]—index of the tail (location from which data will normally be read) for each queue; type **U8**; only exists if at least one queue is configured.

ROM

Two data structures may exist that support queues:

NUSE_Queue_Data[]—contains the start addresses of the memory areas allocated to each queue; type **ADDR**; only exists if at least one queue is configured.

NUSE_Queue_Size[]—size of each queue; type **U8**; only exists if at least one queue is configured.

Pipes

Data structures appertaining to pipes only exist if one or more pipes are configured.

RAM

Three or four RAM-based data structures may exist that support pipes:

NUSE_Pipe_Blocking_Count[]—count of how many tasks are suspended (blocked) on each pipe pending availability of space to write data or the arrival of incoming data; type **U8**; only exists if task blocking is enabled and at least one pipe is configured.

NUSE_Pipe_Head[]—index of the head (location to which data will normally be written) for each pipe; type **U8**; only exists if at least one pipe is configured.

NUSE_Pipe_Items[]—count of the number of data items stored in each pipe; type **U8**; only exists if at least one pipe is configured.

NUSE_Pipe_Tail[]—index of the tail (location from which data will normally be read) for each pipe; type **U8**; only exists if at least one pipe is configured.

ROM

Three ROM-based data structures may exist that support pipes:

`NUSE_Pipe_Data[]`—contains the start addresses of the memory areas allocated to each pipe; type `ADDR`; only exists if at least one pipe is configured.

`NUSE_Pipe_Message_Size[]`—size of messages (in bytes) handled by each pipe; type `U8`; only exists if at least one pipe is configured.

`NUSE_Pipe_Size[]`—size of each pipe; type `U8`; only exists if at least one pipe is configured.

Semaphores

Data structures appertaining to semaphores only exist if one or more semaphores are configured.

RAM

One or two RAM-based data structures may exist that support semaphores:

`NUSE_Semaphore_Blocking_Count[]`—count of how many tasks are suspended (blocked) on each semaphore pending an obtain request; type `U8`; only exists if task blocking is enabled and at least one semaphore is configured.

`NUSE_Semaphore_Counter[]`—counters for each semaphore; type `U8`; only exists if at least one semaphore is configured.

ROM

A single ROM-based data structure may exist that supports semaphores:

`NUSE_Semaphore_Initial_Value[]`—initial count values for each semaphore; type `U8`; only exists if at least one semaphore is configured.

Event groups

Data structures appertaining to event groups only exist if one or more event groups are configured.

RAM

One or two RAM-based data structures may exist that support event groups:

NUSE_Event_Group_Blocking_Count[]—count of how many tasks are suspended (blocked) on each event group pending a specific event or multiple events; type **U8**; only exists if task blocking is enabled and at least one event group is configured.

NUSE_Event_Group_Data[]—data for each event group; type **U8**; only exists if at least one event group is configured.

ROM

No ROM-based data structures exist that support event groups.

Timers

Data structures appertaining to application timers only exist if one or more timers are configured.

RAM

Three RAM-based data structures may exist that support timers:

NUSE_Timer_Expirations_Counter[]—counters of expirations of each timer since the system was initialized; type **U8**; only exists if at least one timer is configured.

NUSE_Timer_Status[]—status of each counter (**TRUE** indicates the counter is running; **FALSE** indicates that it is not); type **U8**; only exists if at least one timer is configured.

NUSE_Timer_Value[]—downcounters for each timer; type **U16**; only exists if at least one timer is configured.

ROM

Two or four ROM-based data structures may exist that support timers:

NUSE_Timer_Expiration_Routine_Address[]—start address of the code for each timer's expiration routine (i.e., a pointer to the function); type **ADDR**; only exists if timer expiration routine support is enabled and at least one timer is configured.

`NUSE_Timer_Expiration_Routine_Parameter[]`—expiration routine parameters for each timer; type `u8`; only exists if timer expiration routine support is enabled and at least one timer is configured.

`NUSE_Timer_Initial_Time[]`—initial time values for each timer; type `u16`; only exists if at least one timer is configured.

`NUSE_Timer_Reschedule_Time[]`—restart time values for each timer; type `u16`; only exists if at least one timer is configured.

Obtaining Nucleus SE source code

Throughout this book, all the source code for Nucleus SE has been included in a piecemeal fashion as it was relevant to the topic in hand. All the source code is available, free of charge, from https://github.com/colin-walls/NUSE. A "getting started" guide is included to help you find your way around the files.

Index

Note: Page numbers followed by "*f*" refer to figures.

Printed in the United States
By Bookmasters